A CITIZEN'S GUIDE TO PRESIDENTIAL NOMINATIONS

Using comparative data on candidate viability from 1976 to 2012, Wayne Steger demonstrates which presidential nominations are effectively won during the invisible primary stage, through the informal coordination of the party elite, and which nominations are determined by voter support in the primaries and caucuses. Through this analysis, Steger illuminates how presidential nomination politics has changed over recent decades and across the two parties.

Barbara Norrander, *Professor of Political Science, University of Arizona*

Presidential nominations in the United States can sometimes seem like a media circus, over-hyped and overly speculative. Even informed citizens might be tempted to tune them out. Yet understanding the process, one distinct to American politics, is crucial for civic participation. If presidential elections are about who will lead the nation, presidential nominations are about who appears on the ballot. This concise and coherent *Citizen's Guide* examines who has power in presidential nominations and how this affects who we as citizens choose to nominate, and ultimately to sit in the Oval Office.

Political scientist Wayne P. Steger defines the nominating system as a tension between an "insider game" and an "outsider game." He explains how candidates must appeal to a broad spectrum of elected and party officials, political activists, and aligned groups in order to form a winning coalition within their party, which changes over time. Either these party insiders unify early behind a candidate, effectively deciding the nominee before anyone casts a vote, or they are divided and the nomination is determined by citizens voting in the caucuses and primaries. Steger portrays how shifts in party unity and the participation of core party constituencies affect the options presented to voters. Amidst all this, the candidate still matters. Primaries with one strong candidate look much different than those with a field of weaker ones. By clearly addressing the key issues, past and present, of presidential nominations, Steger's guide is informative, relevant, and accessible for students and general readers alike.

Wayne P. Steger is Professor of Political Science at DePaul University. He is a former editor of the *Journal of Political Marketing* and co-editor of *Campaigns and Political Marketing*. His work includes studies on agenda-setting, media coverage of campaigns, and marketing in political campaigns. His recent work focuses on party elites, coalition formation, and candidate decision-making.

CITIZEN GUIDES TO POLITICS AND PUBLIC AFFAIRS
Morgan Marietta and Bert Rockman, Series Editors

Each book in this series is framed around a significant but not well-understood subject that is integral to citizens'—both students and the general public—full understanding of politics and participation in public affairs. In accessible language, these titles provide readers with the tools for understanding the root issues in political life. Individual volumes are brief and engaging, written in short, readable chapters without extensive citations or footnoting. Together they are part of an essential library to equip us all for fuller engagement with the issues of our times.

Titles in the series:

A CITIZEN'S GUIDE TO PRESIDENTIAL NOMINATIONS

The Competition for Leadership

Wayne P. Steger

Routledge
Taylor & Francis Group

NEW YORK AND LONDON

First published 2015
by Routledge
711 Third Avenue, New York, NY 10017

and by Routledge
2 Park Square, Milton Park, Abingdon, Oxon, OX14 4RN

Routledge is an imprint of the Taylor & Francis Group, an informa business

Library of Congress Cataloging in Publication Data
Steger, Wayne P.
    A citizen's guide to presidential nominations : the competition for leadership / Wayne P. Steger.
        pages cm. — (Citizen guides to politics and public affairs)
    Includes index.
    1. Presidents—United States—Nomination.   2. Primaries—United States.   I. Title.
    JK521.S74 2015
    324.273'15—dc23
    2014040229

ISBN: 978-0-415-82758-4 (hbk)
ISBN: 978-0-415-82759-1 (pbk)
ISBN: 978-0-203-52248-6 (ebk)

Typeset in Garamond
by Apex CoVantage, LLC

Printed and bound in the United States of America by Publishers Graphics, LLC on sustainably sourced paper.

FOR
LYDIA, CAMILLA, AND LUCCA

CONTENTS

FIGURES

TABLES

SERIES FOREWORD

Political leadership in America depends not only on whether the Commander in Chief is a Democrat or Republican, but *which* Democrat or *which* Republican. In this sense many of our most important political choices are decided at the nomination phase long before the final presidential vote. Wayne P. Steger—one of the foremost scholars of the unique American process of candidate selection—argues that the process can be understood grounded in one overarching question: *Who controls the nomination?* Is it a democratic process determined by the broad membership of each party, or is it decided much earlier and less publicly by party insiders and activists? The answer to this question can change each time the process unfolds. Whether any given contest will be more or less democratic depends on factors clearly explained by Steger in this *Citizen's Guide to Presidential Nominations*, one of the more dynamic and least understood aspects of American politics.

Unlike our general elections, conducted every fourth November under relatively clear rules and traditions, the nomination process is less stable or predictable. The timing, the rules, the financing, the form of voting (in other words, the control of the outcome) have all varied tremendously over time. For reasons explained in this volume, the U.S. Constitution is remarkably silent on the topic of parties and nominations. The process has been the target of several reform movements, most notably in the 1970s when the contemporary nomination process was formed. The replacement of party bosses and convention fights with primary elections and caucuses allowed for the potential for democratic control, but perhaps not its realization. The new system may give power to the voters in early states, who create or deny momentum to specific candidates. Or that power may have been reclaimed by party activists who channel the resources of money and attention to pre-ordained candidates in such a way that their selection is nearly inevitable. This volume explains the short-term and long-term forces that lead each nomination in a specific direction, including the changes in party coalitions, the evolution of campaign financing, and the factors that lead candidates to enter or leave the increasingly long and grueling race. The complex factors determining the nomination have been woven into a clear and concise portrait by Wayne P. Steger, Chair of the Department of Political Science at DePaul University, a former editor of the *Journal of Political Marketing* (2002–2008), and author of many scholarly works on the recent nominations, candidates, campaigns, forecasts, and outcomes. His *Citizen's Guide* offers a clear understanding of how (and by whom) the nomination is controlled, never straying far from the democratic questions at the heart of our system.

Morgan Marietta and Bert Rockman
Series Editors

PREFACE

Presidential elections may be about who will lead the nation, but presidential nominations are about who is on the ballot. Before the 1970s, presidential candidates were selected by party leaders in proverbial smoke-filled rooms, with little input by party members around the country. Reforms of the nomination process in the 1970s were partly successful in bringing more democracy to the selection of the presidential candidates of the two major political parties. But the primaries and caucuses that we have now may or may not have achieved the purpose of handing the control over the nomination to a broad range of American citizens who identify with the major political parties. Party insiders may still hold considerable influence over the selection of presidential nominees if they work together before the caucuses and primaries begin.

Nominations involve both party coalitions and candidates. A nominee is selected when a winning coalition of party constituencies coalesces in support of one of the candidates who aspire to leadership. In some nomination campaigns, party stakeholders—party insiders, activists, and the leaders of aligned groups—come to an early agreement on which candidate should be nominated. In other election years, party stakeholders remain divided or undecided until party voters begin to cast ballots in caucuses and primaries—the nominating elections held in states during the first six months of the election year.

There are important differences in nomination campaigns that are decided *before* or *during* the caucuses and primaries. These are different kinds of nomination campaigns with different processes and different actors exercising power. When nominations are effectively determined before the caucuses and primaries begin, coalition formation occurs through signaling and coalescence among party stakeholders along the lines argued by Marty Cohen, David Karol, Hans Noel, and John Zaller in their book, *The Party Decides*. However, party stakeholders sometimes divide or remain undecided about which candidate to nominate, in which case the formation of a winning coalition within the party shifts to the caucuses and primaries. Coalition formation during the caucuses and primaries occurs as voters learn about candidates who can gain or lose momentum across primaries along the lines argued by John Aldrich in *Before the Conventions*, Larry Bartels in *Presidential Primaries and the Dynamics of Public Choice*, and Samuel Popkin in *The Reasoning Voter*. Most presidential nomination campaigns involve a mix of these scenarios. There usually is some convergence of party stakeholders before the caucuses and primaries but it is not always enough to determine the nominee. This leaves some uncertainty about which candidate will become the nominee as the scope of conflict expands to include the voters in the caucuses and primaries. My argument is that how much party stakeholders are able to coalesce before the caucuses and primaries depends on the cohesion of the party coalitions *and* on the candidates who seek the nomination.

The unity and stability of party coalitions vary over time. Presidential nomination campaigns often involve some degree of intra-party competition because the political parties nationally are more diverse than they are at the state and especially at the local level. Republicans in New York or Massachusetts, for example, are generally more moderate on economic and social policy compared to Republicans from Texas or Oklahoma. The size and diversity of the national political parties makes it harder for party stakeholders to converge and collude in presidential nominations compared to nominations for state, county, or local offices. Further, the stability of the party coalitions differs across time and each party has their own historical path of coalitional formation and fragmentation. For example, Democrats were more divided in the 1960s and 1970s than they have been in the past decade, while the Republican Party is showing signs of internal divisions in the nomination campaigns of 2008 and 2012. These differences show up in the competition for presidential nominations. Unified political parties more readily come together in support of a candidate, while parties that have internal divisions struggle over the selection of their nominee.

Nominations also depend on who runs. Political parties select their nominee, but they have to choose among the candidates who seek the nomination. Candidates differ in their ability to unify the party because they differ in their ideological profiles, political reputations, and personal characteristics like charisma, competence, integrity, and authenticity. Some candidates have more appeal to party constituencies than

others. George W. Bush, for example, was a more appealing candidate
to more of the various constituencies of the Republican Party in 2000
than Mitt Romney was in 2012. While all candidates think they would
make a great president, the various constituencies of their political
party need to be convinced before a winning coalition will emerge.
The degree to which party stakeholders unify early can be determined by looking at how competitive the race is when the caucuses
and primaries begin. When party stakeholders coalesce early, there is
a clear leader in endorsements, public opinion polls, campaign funds,
and news coverage. When party insiders and activists fail to unite early,
there is more competitive balance among the various candidates seeking the nomination. The competitiveness of the caucuses and primaries
also depends on how quickly caucus and primary voters unify behind
a candidate. The caucuses and primaries will be less competitive when
stakeholders coalesce behind a candidate during the invisible primary,
or when caucus and primary voters quickly coalesce behind a candidate
who gains momentum during the caucus and primary season.
    The questions addressed in this book relate to a transcending con-
cern with who exercises power in the selection of presidential candidates. Competition among political parties and candidates is what
enables citizens to make meaningful choices in elections. When a particular candidate has a clear advantage—because party stakeholders have
unified behind that candidate—then there is less competition among
candidates during the caucuses and primaries. When party insiders and
activists unify early, power is exercised by those leaders while voters in
the caucuses and primaries have less meaningful choices. When party
stakeholders remain divided or undecided at the time the first decisions
are made in the Iowa Caucus, then there will be more viable candidates for caucus and primary voters to select among. Power effectively
shifts to voters in presidential caucuses and primaries whose decisions
determine the nominee. Thus the degree to which presidential nominations are competitive at the onset of the caucus and primary informs
us about how democratic the selection process is and who holds power
in the American political system.

ACKNOWLEDGMENTS

My interest in presidential nominations originated in a course on the American presidency at Marquette University. One astute student, Ryan McAlvey, questioned why campaign momentum didn't seem to have the explanatory power posited by the literature at the time. Our conversations sparked my curiosity and led me to analyze news coverage of nomination campaigns, a line of inquiry that has since expanded and evolved along the way. Randall Adkins and Andy Dowdle have greatly influenced my thinking about presidential nominations through our conversations and collaborations, particularly with respect to candidate ambition and campaign fundraising. Randy's insights about candidate ambition have been particularly important for the second half of the story of this book. Lara Brown has underscored the point that candidates are opportunists as well as ambitious. Dante Scala and Andrew Smith have been helpful resources with respect to the role of early caucuses and primaries. Barbara Norrander has provided a great deal of insight about the process of party nominations, particularly as a factor framing candidate decision-making. Christopher Hull has emphasized to me the importance of candidate viability as a factor in voters' decision-making. Andrew Dowdle and Andrew Smith graciously provided feedback on an early draft.

Seth Masket, Hans Noel, and David Karol have pushed me to consider arguments about the growing role of political party activists. The political parties are structurally and operationally different from what they were before the 1970s. The proliferation of quasi-independent groups with intense demands for policy has had a profound impact on the political parties and presidential nominations. My father, Charles Steger, is a reminder to me of the potency of these groups to influence the political preferences of active partisans. While political party activists are intense policy demanders, the coalitions are not permanent.

Arthur Paulson influenced my thinking about the variability of party coalition unity as a factor in nominations. Bill Mayer and Zach Cook have influenced my thinking about public opinion and party constituencies. In all cases, the book is better through my interactions with these folks and any errors are mine alone.

I also want to thank Morgan Marietta who provided feedback on the thematic coherence and writing of the book. Michael Kerns has been a great editor and a pleasure to work with.

Finally, I dedicate the book to my children—Lydia, Camilla, and Lucca.

Part I

INTRODUCTION AND RULES
OF THE GAME

1

AN INTRODUCTION TO PRESIDENTIAL NOMINATIONS

When citizens vote in elections, they choose between the candidates nominated by the two major political parties. Other candidates may appear on the ballot, but only the nominees of the Democratic and Republican Parties have a realistic chance of being elected. Although citizens may wish that they had some "other" candidate they could vote for, they are effectively limited to the candidates nominated by the major parties. The ability to put candidates on the ballot makes political parties the arbiters of representative democracy.[1] As arbiters, the parties have the capacity to choose candidates that serve partisan interests, potentially at the expense of the broader public.[2]

Nominating candidates is one of the main functions of political parties, which exist mainly to coordinate the efforts of groups that join together in order to win elections and, in turn, to control government for the purpose of delivering policies that are favorable to their interests. The selection of nominees is critically important to the political parties. The presidential nominee becomes the foremost spokesperson and the personified image of the party.[3] Presidential nominations also affect the ideological direction of the party.[4] The choice of presidential nominees helps shape what the parties stand for and what they will try to do should they win the election.

How the nominee is selected and who does the selecting also matters for broader questions about representative democracy in America. Historically, presidential candidates were selected by party leaders in proverbial smoke-filled rooms, with little input by citizens. Reforms of the nomination process in the 1970s enabled larger numbers of citizens to participate in the selection of the party nominee in presidential primaries. Yet citizens voting in presidential caucuses and primaries may not have as much choice as it appears.[5] Party insiders may still hold considerable influence over the selection of presidential nominees if they work together before the caucuses and primaries begin. Collusion by party insiders can create an uneven playing field that gives their preferred candidate an easier path to the nomination.

The extent to which citizens have a meaningful choice in caucuses and primaries depends on how competitive the nomination campaign is. Citizens are empowered to select their leaders when they have several viable candidates to choose among in an election. If party stakeholders coordinate their efforts to give a competitive advantage to one candidate before the caucuses and primaries begin, then citizens voting in these party nominating elections may have little choice but to go with the choice of party stakeholders. The vote is little more than a symbolic endorsement of the candidate selected by party stakeholders. If party stakeholders fail to unify in support of a preferred candidate, however, then voters in the caucuses and primaries will have a larger number of viable candidates to choose among. In this case, voters have a more meaningful role in selecting the nominee.

Presidential nominations involve the interactions between party constituencies seeking to satisfy their own preferences and priorities, and opportunistic politicians seeking to advance their own ambitions. Winning a presidential nomination requires assembling a winning coalition of party constituencies in support of a candidate. This coalition-building is an interactive process. Candidates compete for the support of party constituencies, and party constituencies seek a winner who will champion their political and policy priorities. A nominee is selected when a critical mass of party constituencies unify behind a candidate.

Historically, the process of unifying behind a candidate occurred during the national nominating conventions where party bosses negotiated with each other and with representatives of the candidates in exchange for patronage and policies.[6] Over time, the period before the convention became more important as candidates and their supporters sought to secure commitments of delegates in presidential caucuses and primaries.[7] Reforms made during the early 1970s moved the formal selection of presidential nominees from the conventions to the state caucuses and primaries that select convention delegates pledged to candidates.[8] During the 1980s and 1990s, however, subsequent modifications to the rules and increasing coordination among party stakeholders enabled winning coalitions to form even before the state caucuses and primaries begin.[9]

Political party coalitions consist of party stakeholders and party identifiers. Party stakeholders are the politicians, party officials, leading party activists, and leaders of groups that are aligned with a party. They

constitute an elite subset of the party membership. Party identifiers are
the citizens who think of themselves as Democrats or Republicans and
who generally vote for their party's candidates in elections. Most of the
people who vote in caucuses and primaries are party identifiers. If party
stakeholders can unify in support of a candidate before the caucuses
and primaries begin, the tendency is for party identifiers to support
that candidate once the caucuses and primaries are underway. If party
stakeholders remain divided or undecided, however, then citizens voting in caucuses and primaries tend to base their decisions less on the
cues provided by party stakeholders and more on things like candidate
appeal and the results of early caucuses and primaries.

The 2000 Republican presidential nomination campaign exemplifies
the pattern of party stakeholder coalescence before the caucuses and
primaries. George W. Bush emerged as the favorite of party stakeholders during the invisible primary—the early phase of the campaign that
occurs before the Iowa caucus and the New Hampshire primary (the
first nominating elections that select state delegates to the national party
conventions). Bush received the most endorsements from party elites
and group leaders. He raised the most money and received the most
news coverage. He consistently received a majority in national public
opinion polls of Republican Party identifiers in 1999. The Republican
caucuses and primaries of 2000 largely confirmed the decisions made
earlier by party stakeholders.

In other years, however, party stakeholders have failed to coalesce
sufficiently to enable their preferred candidate to win the nomination.
In such years, party stakeholders either divide their support among the
candidates or they remain uncommitted until the caucuses and primaries begin.[10] The 2008 Democratic race illustrates this kind of presidential nomination. Senator Hillary Clinton was the front-runner during
the invisible primary with more endorsements, money, and support
in polls than any of her rivals, including Senator Barack Obama. But
Clinton was not as strong a front-runner as Bush had been eight years
earlier. The majority of party insiders did not endorse her candidacy.
In fact, most refrained from making an endorsement or they endorsed
other candidates. Clinton's support in national opinion polls consistently fell short of a majority. By 2008, the internet had obliterated
the traditional fundraising advantage of front-runners since candidates
could raise large sums of money quickly online. The invisible primary
of 2007 thus did not yield a winning coalition of Democrats behind
Clinton's candidacy. Instead, Obama's victory in the Iowa caucus gave
him substantial momentum and Democratic voters moved to support

5

him in the caucuses and primaries that followed.[11] The critical decisions were made by the citizens who voted in the state caucuses and primaries held during the first six months of the election year.

These two presidential nominations reflect differences in intra-party coalition coalescence and differences in who exercised power in selecting the presidential nominees. Nominations that are effectively decided *before* the caucuses and primaries reflect an insider game in which party stakeholders play the dominant role in the selection of the nominees. When stakeholders fail to unify sufficiently, the critical period of coalition formation shifts to the caucuses and primaries, and campaign momentum becomes a more potent factor in determining the nominee. Understanding which pattern will occur requires figuring out why party stakeholders are able to unify behind a front-runner more in some years than they do in other years.

Coalition Coalescence Before and During the Primary Season

The reforms of the early 1970s moved the formal selection of the presidential nominees from the national nominating conventions to the state caucuses and primaries. Caucuses are meetings run by state and local party organizations at which party members meet to discuss candidates and issues and cast ballots for the party's candidates. These nominating elections select delegates to county or state conventions that formally select delegates to the national conventions. Primaries are elections run by state and local governments in which eligible voters cast ballots for the party's candidates. The results of a primary election determine the delegates to the national conventions. Eligibility varies by state, but the voters in most state caucuses and primaries are citizens who identify themselves as Democrats or Republicans. While caucus and primary voters determine the candidates' delegates to the national convention, the nomination campaign begins much earlier, during what is commonly called the invisible primary—the early phase of the campaign occurring during the year or so prior to the election year when candidates and party stakeholders are engaged in an interactive search for leadership.

The invisible primary has been the critical period of coalition coalescence in support of a nominee in most presidential nominations since the 1970s. During this phase of the campaign candidates reach out to party stakeholders to gauge their potential backing and get pledges of support. Party stakeholders evaluate candidates and try to figure out which candidates are preferable, viable, and electable. Candidates are preferable mostly in terms of their personal characteristics and their positions on policies demanded by party stakeholders. Candidates are viable if they are perceived as being able to win the nomination—that is that they will be able to get support from the various constituencies of the party. Party stakeholders try to figure out which candidate they prefer and whether that candidate is preferred by other stakeholders as well as by party identifiers. Thus, candidates are deemed preferable and viable mainly in terms of their appeal to party constituencies. Candidate electability refers to perceptions that a candidate can win a majority in the general election *if* they are selected as the party's nominee. Electability depends not only on a candidate's appeal to party constituencies but also to other citizens who vote in the general election. Some candidates may be preferred by party constituencies but may not be electable. Others may be electable but are not preferred by party constituencies.

The argument that the invisible primary is the critical period in the nomination campaign essentially holds that a winning nominating coalition forms before the caucuses and primaries. Marty Cohen, David Karol, Hans Noel, and John Zaller have described the invisible primary as a "long national discussion" among party insiders, activists and groups who send signals to each other about which candidate they believe will be able to promote party policy positions *and* can win nomination and the general election.[12] Party stakeholders are able to exert influence over presidential nominations in part because most people do not pay a lot of attention to politics and to the claims made by the candidates.[13] Party stakeholders are intensely concerned with politics and policy so they have the incentive to attempt to influence the selection of the nominee. They want a nominee who will deliver benefits to members of the party coalition. Stakeholders try to figure out which candidate is most able to win and deliver those benefits, and they try to rally around that candidate even before the caucuses and primaries begin. If these stakeholders come to agreement on which candidate should be nominated, then they can collude by massing their support behind a single candidate who becomes highly likely to become the party's nominee.[14] Although party identifiers still vote in the presidential caucuses and primaries, the tendency is for party identifiers to follow the lead of party stakeholders and support that candidate.

Party stakeholders, however, do not always unify behind a candidate enough to enable this candidate to win the nomination. While the media always label one candidate as the front-runner during every campaign, front-runners differ in the size of their advantage going into the primaries. Some front-runners are strong enough to prevail in the primaries and caucuses like George W. Bush in 2000. Other so-called front-runners lack enough support from party stakeholders to secure the nomination, as Hillary Clinton found out in 2008. One reason is that party stakeholders may divide among themselves with various stakeholders supporting different candidates. More often, party stakeholders may be unsure about which candidate is the most viable, so they take a wait-and-see approach to see which candidate generates enthusiasm among party constituencies. When party stakeholders divide or remain undecided, they may not be able to muster sufficient support in time to help a candidate to secure the nomination during the caucuses and primaries.

When party stakeholders are divided or undecided, they deprive rank-and-file partisans a clear signal about which candidate should be supported. Uncertainty about which candidate will win the nomination affects the strategies and behaviors of candidates, party insiders, campaign contributors, the media, and the citizens who vote in caucuses and primaries. In these races, no candidate secures the majority of endorsements by party stakeholders. Candidates are more evenly matched in the amount of money they raise so there is more parity in their ability to communicate with prospective voters. The news media divide their coverage more evenly among candidates so more candidates become known to prospective voters. Public opinion surveys indicate that the front-runners have the support of only a plurality of the party identifiers responding to these polls. These nominations are relatively competitive when the caucuses and primaries begin.

If no candidate obtains the support of a majority of party stakeholders during the invisible primary, the critical period of coalition formation shifts to the caucuses and primaries. The process of intra-party coalition formation changes with respect to participants and dynamics. Party stakeholders are joined by more of the mass membership of the political parties who vote in the caucuses and primaries. The expansion of the scope of competition adds uncertainty to the race, as increasing numbers of party voters begin to pay closer attention to the candidates. The influence of the cues given by party stakeholder support diminishes while the results of early caucuses and primaries become more important. The results of the early caucuses and primaries affect candidate behavior, media coverage, fundraising, and voting in the subsequent caucuses and primaries. In these circumstances, "campaign momentum" during the caucuses and primaries can play a big role in determining the nominee.[15] Campaign momentum is the change in a candidate's position in a caucus or primary relative to the candidate's pre-contest standing. In a "momentum" campaign, support for candidates can shift from primary to primary. One or two candidates emerge at the top while other candidates drop out of the race. Campaign momentum reflects a dynamic process in which a candidate "beats expectations" in a presidential caucus or primary and gains momentum while candidates that finish below expectations lose momentum. Expectations are set by the political commentators, journalists, and others who comment on the campaign in the media. Candidates try to influence these expectations— downplaying their own chances so that they look good while trying to set up their rivals for failure by raising expectations for these rivals. Candidates that beat expectations in a caucus or primary usually experience a surge in media exposure, fundraising, and support in public opinion polls. These candidates are perceived as more viable, which can change the candidate preferences of voters casting ballots in subsequent caucuses and primaries.[16]

At the same time, there is increasing pressure on "losing" candidates to abandon the race and help unify the party. Candidates who fail to meet expectations typically receive less media coverage and the coverage that they do receive is more critical of their candidacy. These candidates also are less able to raise money, which means that they have even less ability to campaign in subsequent caucuses and primaries. Candidates who lack campaign funds usually quit the race.[17] The nominee is determined as support for one candidate builds while other candidates withdraw. The caucuses and primaries occurring later in the schedule often are less competitive because most if not all of the rival candidates have dropped out of the race.

It is important to realize that these two perspectives are not incompatible and both exist to some extent in every election because party stakeholders vary in the extent to which they unify behind a candidate before the caucuses and primaries. There usually is some convergence of party stakeholders before the caucuses and primaries but it is not always enough to determine the nominee. This leaves some uncertainty about which candidate will become the nominee as the scope of conflict expands to include the voters in the caucuses and primaries. Once the caucuses and primaries begin, the signals sent by party stakeholder endorsements become less important than the results of these contests.

Variation in when and to what extent the parties unify before the caucuses and primaries creates differences in *how nominees are selected*, even though the rules for electing delegates to the national nomination campaigns have remained largely the same since 1980. If party stake-holders come to agreement on a candidate before the caucuses, then the nominee is essentially selected in an "insider" game through a process of signaling and coordination among these players. If stakeholders fail to unify sufficiently, then the nominee is essentially selected in an "outsider" game in which citizens participating in the caucuses and primaries select the nominee. These two patterns of presidential nomination campaigns differ in *who exercises power*. When presidential nominations are effectively decided *before* the caucuses and primaries, then we can infer that party stakeholders played a dominant role in the selection of the nominees. When the competition of the invisible primary is indeterminate, the nominee is effectively decided *during* the caucuses and primaries, and we can infer that the participating mass membership of a political party played the dominant role in selecting the nominee.

Which Kind of Nomination Campaign Will Occur?

Understanding why one pattern or the other occurs in a given election requires figuring out why party stakeholders unify or coalesce behind a candidate more in some years than in others. The argument of this book is that whether the critical period of coalition coalescence occurs during the invisible primary or during the caucuses and primaries depends mainly on the *cohesion of the party coalitions* and on which *candidates enter the race*.

The party side of the explanation requires recognition of two aspects of party coalitions. First, the political parties are diverse coalitions of groups and constituencies, more so at the national level than at the state or local level. As the diversity of party coalitions increases, it becomes harder for party stakeholders to reach agreement on the choice of a nominee. The diversity of the national political parties makes it harder for party stakeholders to collude in presidential nominations compared to nominations for state, county, or local offices. Second, the cohesiveness of the party coalitions varies across time. The party coalitions evolve, with each party having their own historical path of coalition formation and fragmentation. The party coalitions that nominate presidential candidates change as the political parties realign (a long-term effect) and as participation varies across nomination campaigns (a short-term effect). Cohesive political parties can more readily come together in support of a candidate, while parties with internal divisions will have more competitive nominations.

10

The major political parties are broad coalitions of groups and constituencies in the American electorate. There are multiple possible winning coalitions that can be formed in any broad collection of people
and groups like a national political party.[18] As political parties become
more diverse, there are more possible combinations of constituencies
that can be assembled to win a majority in the nominating process.
The national parties consist of a more diverse collection of groups
and constituencies than exist at the state level. While diversity adds to
the innovativeness and cultural richness of America, it also gives rise
to differences of interests, beliefs, and opinions that result in political
conflict. Disagreements among the stakeholders of a political party can
result in competition over nominations. This is why factions of a political party sometimes fight over nominations, as illustrated by the recent
clashes between groups affiliated with the Tea Party movement and
"establishment" Republicans over nominations for congressional elections. While party stakeholders have incentives to coordinate among
themselves to select a presidential nominee, the diversity of the political
parties at the national level makes it harder to collude in the selection
of nominees compared to nominations for congressional or state-level
offices.

Party coalitions also change over time. One kind of change occurs as
the political parties realign (a long-term effect). Political party realign-
ment usually involves a slow change in the membership composition
of the political parties.[19] The constituencies of the Democratic and
Republican parties have changed substantially over the past fifty years.
In particular, Southern white voters have gravitated away from the
Democratic Party and toward the Republican Party, while voters in
Northeastern states are less Republican Party and more Democratic.
These long-term changes in the memberships of the political parties
affect the formation of nominating coalitions within each of the political parties. It is easier for party stakeholders to reach agreement on
presidential nominees when the party coalitions are stable compared to
when they are in the process of realigning to a different configuration
of groups and constituencies.

For example, the Democratic Party coalition that emerged during
the "New Deal" of Franklin D. Roosevelt started to fragment beginning in 1938 and increasingly in subsequent years. The Democratic
Party remained relatively divided from the 1960s into the 1990s though
the party's coalitional unity has been increasing as more conservative
members have left the party.[20] The Republican Party, by contrast, was
relatively divided during the 1950s to the 1970s but gained unity during

the Reagan years. But even the relatively unified Republican Party coalition has internal divisions over social issues and the relative primacy of balanced budgets or tax cuts.[21] These kinds of internal divisions in a political party make it harder to find a candidate whose appeal unifies the various factions. Intra-party divisions may emerge as episodic responses to the issue environment in a given election cycle. These sources of variation in coalition unity also give candidates some opportunity to develop strategy and messages that will attract a winning coalition of party constituencies during the nomination campaign.

Party constituencies also vary in how much they participate in a given presidential nomination campaign (a short-term effect). Political activism is costly in terms of time and money. Party activists and groups may get involved in some nomination campaigns while sitting on the sidelines in other years. Howard Dean's supporters in the 2004 Democratic nomination, for example, consisted of a good many anti-war activists—many of whom had opposed the Vietnam War but who had not been particularly active in presidential nominations until the War in Iraq.[22] While party coalitions are largely the same from election to election, there is some variation in the engagement of different party constituencies. Participation may vary as a result of changing agendas and efforts by candidates to mobilize support for their nomination campaign. Barack Obama, for example, had greater ability than his rivals to motivate younger voters and African Americans to participate in Democratic presidential caucuses and primaries. Greater participation by a specific constituency group can make a difference for the outcome in a close nomination campaign.

It is easier for stakeholders to coalesce around a candidate when a political party coalition is relatively stable and unified. It is harder for party stakeholders to come to agreement on a candidate when a party is internally divided among factions that may not share the same policy priorities or when the membership of a party is changing.

Whether presidential nominations are settled earlier by party stakeholders during the invisible primary or later by voters in the caucuses and primaries also depends on who enters the race. "Who runs" matters in a presidential nomination campaign. Every candidate brings a unique package of personal characteristics, policy positions, and ideological image to a campaign. Some candidates will have broader appeal among the different constituencies that make up the political parties than do other candidates. While all candidates think that they have the right stuff to be president, their self-image and policy vision may not be shared by all segments of their political party. Politicians like Ron Paul

(TX) and his son Senator Rand Paul (KY), for example, believe that the Republican Party—and America more generally—would be better off following their vision of limited government across the board for economic, social, and foreign policy. Only a minority of Republicans, however, share that libertarian vision on policy priorities like national defense and gay marriage.

Candidates who have broader personal and policy appeal have an easier time attracting support from the various constituency groups of the party compared to candidates who have narrower—but potentially stronger—appeal to specific constituencies of a political party. Even candidates who try to make broad appeals, like Mitt Romney, may have difficulty attracting the support of some political party constituencies that may have doubts about a candidate's personal character or their commitment to certain policy positions. Romney, for example, rarely received more than two-thirds of the vote in Republican primaries— even after the other major candidates had withdrawn from the race. Conservatives may have had doubts about Mitt Romney's commitment to their principles during the primaries, but they supported him in the general election when the alternative was a Democrat.

A candidate's appeal depends in part on the values, beliefs, and preferences of the people who are evaluating them. Who participates in the selection process determines the set of beliefs and preferences that are relevant to the selection of the nominee. Who participates in presidential nominations has changed over time, as will be discussed in later chapters. But an important consideration to remember is that the participants in the nomination process select among candidates who have their own unique set of personal, political, and policy characteristics. *Personal characteristics* like integrity, charisma, competence, leadership, and authenticity affect citizens' preferences for candidates. *Political characteristics* refer to a candidate's reputation and position within the space of a political party. *Policy characteristics* are similar, but refer to a candidate's ideological and policy promises. Candidates for a political party nomination often adopt similar policy stances on policies that are important to party constituencies, but they may differ in their priorities and credibility. Party activists want to know that a candidate is committed to their issue concerns. While candidates say they are committed to all of their positions, party activists want to be sure the candidate can be trusted to fulfill their promises. Doubts about policy commitments can soften a candidate's support among party activists and group leaders.

The people who participate in presidential nominations are mainly party stakeholders and party identifiers who have strong preferences for policy. Candidates promise to deliver those policies. A candidate's personal and political characteristics give clues about a candidate's capacity to deliver on those promises should the candidate be elected. A candidate's friends within a political party's constellation of factions and interests provide additional information about a candidate's ability and credibility to deliver on promises. The campaign itself matters because candidates have at their disposal—if they can raise the money to pay for it—a wide range of marketing tools and techniques that can be used to maximize their appeal to voters. Voters' preferences for candidates can be manipulated to some extent by campaign appeals that activate certain beliefs and feelings rather than other beliefs.[23]

Nomination campaigns are complex. Understanding them requires thinking about the interaction of the party constituencies that participate in the selection process and the candidates who decide to run. The unity and stability of party coalitions changes over time, introducing variation in the ability of party stakeholders to reach an early agreement on the choice of nominees. The ability to unify also depends on which candidates run since some candidates have greater appeal than others. The degree to which party stakeholders unify early can be determined by looking at how competitive the race is when the caucuses and primaries begin. When party stakeholders coalesce early, there is a clear leader in endorsements and in public opinion polls. When party insiders and activists fail to unite early, however, there is more balance in the resources of the candidates seeking the nomination and campaign momentum during the caucuses and primaries has a greater impact on the outcome.

The Relevance to Representative Democracy

How leaders are selected and who participates in the selection of leaders informs us about how democratic presidential nominations are. Having more people involved in the selection of presidential nominees is consistent with representative democracy, which holds that power should derive from the people directly or indirectly. Presidential nominations can be said to be less democratic as party stakeholders exert more influence over the selection of the nominees, and more democratic as the choice depends on the decisions of caucus and primary voters.

In one version of democratic theory, it is competition among political organizations and leaders that provides citizens with the opportunity to make meaningful choices in elections.[24] When a particular candidate has a huge lead because party stakeholders have unified behind that candidate, then there is less effective competition among candidates during the caucuses and primaries.[25] In a campaign in which resources and insider support are concentrated behind one candidate, primary and caucus voters basically have a vote of confidence (or no confidence) in the candidate preferred by party stakeholders. Even then, voters will have a hard time finding a viable alternative to the front-runner if party stakeholders are unified. However, if party stakeholders are undecided or divide their support among different candidates, then voters become the arbiters of the nomination competition. Presidential nomination campaigns can be considered more democratic when caucus and primary voters are able to exert more power over the selection of the eventual nominee.

Analyzing the competitiveness of presidential nomination campaigns enables us to identify the extent to which party stakeholders unify in support of a candidate, when coalition coalescence occurs, and thus what it means for democracy in the selection of presidential candidates. Nominations in which party stakeholders are able to unify by the end of the invisible primary are less competitive once the voting begins and thus tend to be less democratic. Nominations in which party stakeholders fail to unify tend to have several viable options for voters in the caucuses and primaries, effectively giving those voters more influence over the selection of the nominee. A more democratic nomination process, however, may not produce better outcomes for voters in the general election—a question that we will revisit in the final chapter.

Notes

1 E. E. Schattschneider, 1960, *The Semi-Sovereign People*, New York: Harcourt Brace.
2 Kathleen Bawn, Martin Cohen, David Karol, Hans Noel, Seth Masket, and John Zaller, 2012, "A Theory of Parties: Groups, Policy Demanders and Nominations in American Politics," *Perspectives on Politics*, 10(3): 571–597; Jeffrey M, Stonecash, 2013, *Understanding American Political Parties, Democratic Ideals, Political Uncertainty, and Strategic Positioning*, New York: Routledge.
3 Arthur H. Miller and Bruce E. Gronbeck, 1994, *Presidential Campaigns and American Self Images*, Boulder, CO: Westview.
4 William H. Riker, 1986, *The Art of Political Manipulation*, New Haven, CT: Yale University Press; Stephen Skowronek, 1993, *The Politics Presidents Make: Leadership from John Adams to George Bush*, Cambridge, MA: Belknap Press; Richard Herrera, 1995, "The Crosswinds of Change: Sources of Change in the Democratic and Republican Parties," *Political Research Quarterly*, 48: 291–312; David Karol, 2009, *Party Position Change in American Politics: Coalition Management*, New York: Cambridge University Press.

5 Wayne Steger, 2000, "Do Primary Voters Draw from a Stacked Deck? Presidential Nominations in an Era of Candidate-Centered Campaigns," *Presidential Studies Quarterly*, 30(4): 727–753.
6 Paul T. David, Ralph M. Goldman, and Richard C. Bain, 1960, *The Politics of National Party Conventions*, Washington DC: Brookings Institution; Eugene Roseboom, 1970, *A History of Presidential Elections*, New York: Macmillan.
7 William R. Keech and Donald R. Matthews, 1976, *The Party's Choice*, Washington DC: Brookings Institute; Howard L. Reiter, 1985, *Selecting the President: the Nominating Process in Transition*, Philadelphia: University of Pennsylvania Press.
8 John Aldrich, 1980, *Before the Convention: Strategies and Choices in Presidential Nominations*, Chicago, IL: University of Chicago Press; Larry M. Bartels, 1988, *Presidential Primaries and the Dynamics of Public Choice*, Princeton, NJ: Princeton University.
9 Marty Cohen, David Karol, Hans Noel, and John Zaller, 2008, *The Party Decides: Presidential Nominations Before and After Reform*, Chicago: University of Chicago Press.
10 Wayne P. Steger, 2008, "Inter-Party Differences in Elite Support for Presidential Nomination Candidates," *American Politics Research*, 36(1): 724–749; Wayne P. Steger, 2013, "Two Paradigms of Presidential Nominations," *Presidential Studies Quarterly*, 43(2): 377–387.
11 Loren Collingwood, Matt A. Barreto, and Todd Donovan, 2012, "Early Primaries, Viability and Changing Preferences for Presidential Candidates," *Presidential Studies Quarterly*, 42(2): 231–255.
12 Cohen et al., 2008, *The Party Decides*.
13 Bawn et al., 2012, "A Theory of Parties"; Kathleen Hall Jamieson, Richard Johnston, and Michael G. Hagen, 2000, *The 2000 Nominating Campaign: Endorsements, Attacks, and Debates*, Research Report, Annenberg Public Policy Center, University of Pennsylvania.
14 Andrew Dowdle, Scott Limbocker, Song Yang, Karen Sebold, and Patrick A. Stewart, 2013, *The Invisible Hands of Political Parties in Presidential Elections: Party Activists and Political Aggregation from 2004 to 2012*, New York: Palgrave Pivot.
15 Larry Bartels, 1985, "Expectations and Preferences in Presidential Nominating Campaigns," *American Political Science Review*, 78(4): 804–815. Candidates talked about momentum and winnowing before these phenomena gained currency in scholarship. Senator Fred Harris characterized the nominating system as a "winnowing process" and his campaign sought to use the Iowa caucus "to boost his chances going into New Hampshire" (quoted on January 20 and March 3, 1976, broadcasts of CBS evening news).
16 Collingwood, Barreto, and Donovan, 2012, "Early Primaries, Viability, and Changing Preferences for Presidential Candidates."
17 Barbara Norrander, 2006, "The Attrition Game: Initial Resources, Initial Contests and the Exit of Candidates During the US Presidential Primary Season," *British Journal of Political Science*, 36: 487–507.
18 William H. Riker, 1962, *A Theory of Political Coalitions*, New Haven, CT: Yale University Press.
19 V. O. Key, 1959, "Secular Realignment and the Party System," *Journal of Politics*, 21(2): 198–210; Walter Dean Burnham, 1970, *Critical Elections and the Mainsprings of American Politics*, New York: W.W. Norton; James L. Sundquist, 1983, *Dynamics of the Party System: Alignment and Realignment of Political Parties in the U.S.*, Washington DC: Brookings Institute.

20 William G. Mayer, 1996, *The Divided Democrats: Ideological Unity, Party Reform, and Presidential Elections*, Boulder, CO: Westview Press.

21 While both fiscal conservatives and supply-side Republicans prefer smaller government and lower taxes, fiscal conservatives give greater priority to balanced budgets while supply-side conservatives prioritize tax cuts. Fiscal conservatives show willingness to raise taxes to reduce budget deficits, while supply-side conservatives have shown willingness to increase budget deficits in order to obtain tax cuts.

22 Scott Keeter, Cary Funk, and Courtney Kennedy, 2005, "Deaniacs and Democrats: Howard Dean's Campaign Activists," Paper presented at the State of the Parties conference, University of Akron, Akron, OH, October 5–7.

23 Larry M. Bartels, 2006, "Priming and Persuasion in Presidential Campaigns," in *Capturing Campaign Effects*, Richard G. C. Johnston and Henry E. Brady (eds.), Ann Arbor: University of Michigan Press, 78–112; James N. Druckman, 2004, "Priming the Vote: Campaign Effects in a US Senate Election," *Political Psychology*, 25(4): 577–594.

24 Joseph Schumpeter, 1942, *Capitalism, Socialism, and Democracy*, London: Allen and Unwin; David Held, 1987, *Models of Democracy*, Stanford, CA: Stanford University Press.

25 Wayne P. Steger, John Hickman, and Ken Yohn, 2002, "Candidate Competition and Attrition in Presidential Primaries, 1912–2000," *American Politics Research*, 30(3): 528–554.

2

AN EVOLVING NOMINATION PROCESS

Presidential nominations are determined by who decides and who runs. Both factors are affected by the rules that structure the interaction of party constituencies and the candidates seeking the nomination. Rules impose constraints on participants and shape the incentives of both sides of the nomination equation—the politicians seeking the nomination and those seeking to influence the choice of nominee. Rules define who can participate in the selection process, which determines whose policy demands will be promoted by a political party. The rules also influence potential candidates when they decide whether or not to seek the nomination.

Because political parties control the selection of candidates—outside of the constraints of the Constitution—the rules of nomination process have evolved with the political parties. The composition, structure, and operation of the political parties tend to change because the parties need to adapt to changing social and economic conditions. Changing social and economic conditions create new issues that may cut across existing party lines, create new constituencies with different demands for policy, and even alter the policy demands of existing party constituencies. Political parties need to adapt and adjust to these changing conditions as they seek to win elections. Changes in the rules and norms of presidential nominations generally reflect changes in the membership of the political parties and the relative power of different groups that form the party coalitions. Rules thus both reflect and affect who has power within a political party.

The process of nominating presidential candidates tends to change when those holding power in a political party need to concede influence over the selection of party nominees in order to retain legitimacy and gain the votes needed to win the general election.[1] The expansion of participation in the nominating process brings new voices into a

18

political party and alters the policies promoted by a party. Such changes can potentially dilute or alter the party's commitment to existing policy positions. Given these potential costs, party stakeholders generally resist opening the nomination process to new participants—doing so only when it is necessary to attract additional supporters in order to win elections. Stakeholders find that it is preferable to share power with new coalition partners rather than lose elections. The general pattern has been one in which changes in the rules have opened the nomination process to more participants, followed by an effort to regain some control of the nomination process. Each period of change in the rules has been followed by a period in which the established power brokers of the prior era have sought to reestablish some of their influence over nominations.[2]

While the political parties adapt and adjust to changing social and economic conditions, the nominating process tends to change significantly in periods of reform rather than gradually over time. There are long periods of time in which the rules of the nomination process are stable with few if any significant changes. These periods are referred to as presidential nomination systems, which are characterized by a particular set of rules and norms that form the framework for selecting presidential nominees. The rules and norms define who can participate in the nomination process, who has influence within the political parties, and whose demands are being served by the nomination process.

The First Presidential Nomination System

The U.S. Constitution is silent on presidential nominations. It specifies only that the president is to be selected by a majority in the Electoral College—an assembly of Electors who are selected by each state to cast the ballots that formally elect a president.[3] Originally most states did not even hold elections in which citizens could vote for the president. Instead, the state legislatures selected the Electors. It was not until 1832 that all of the states had presidential elections in which citizens' votes determined the appointment of Electors to the Electoral College. Whether by popular election or selection by state legislatures, there is no constitutional procedure for determining who would be considered by the states when selecting Electors. This seems odd from a contemporary perspective, but the Framers did not appreciate or anticipate political parties. Instead they believed that the best leaders would be chosen from among the prominent men of the nation.

19

The nomination of presidential candidates was not a pressing issue in the nation's first presidential election. Supporters of the new Constitution sought to build support for the new national government by supporting George Washington to be the first president. Recognizing the appeal of Washington, other potential candidates decided not to run and he was elected by acclamation in the Electoral College. The decision not to run by other potential candidates certainly made Washington's election easier.

During Washington's two terms in office, factions emerged to contest the election of the next president. Groups of legislators in Congress met to nominate candidates and coordinate campaign efforts in their home states. These factions formed the nation's first political parties—the Federalists, who sought greater national government authority, and the Democratic-Republicans or Jeffersonians, who generally sought greater protections for states' rights. The key players in the nominations during this era were the elected officials in Congress and their allies in the state legislatures.

This "congressional caucus" system, however, had several problems. One, it violated the constitutional principle of separation of powers since members of Congress determined who could be selected as the president. Two, the system failed to represent party constituencies in states and congressional districts that elected legislators of another party. Citizens whose preferred party failed to win a congressional election had no voice in the selection of presidential nominees. For example, citizens from Southern states who supported the Federalist Party did not have a voice in the selection of Federalist Party presidential candidates after 1800 because Jefferson's Democratic-Republican Party won congressional elections in those areas.

The congressional caucus ceased to function as a national nominating mechanism in 1824 when the Democratic-Republican Party split into multiple factions in Congress that nominated four candidates in different sections of the country. Under the U.S. Constitution, if no candidate receives a majority in the Electoral College, the president is selected by the newly elected House of Representatives with each state delegation getting one vote. Deal making in the House of Representatives led to the selection of John Quincy Adams, even though Andrew Jackson had received more votes and Electors than any of the other candidates. This outcome threatened the legitimacy of the government.

After losing the 1824 election despite getting more votes than any other candidate, Andrew Jackson and his supporters established local party organizations across much of the country to support Jackson's candidacy in the 1828 presidential election. Power over presidential nominations effectively shifted from centralized decision-making by members of Congress to more decentralized decision-making by state and local party organizations. These party organizations solved the problem revealed by the 1824 election by nominating presidential candidates and coordinating their campaign efforts across states for the general election.

The Second Nomination System: The Caucus-Convention System

Beginning with the presidential election of 1832, the nation's major political parties have held national conventions to formally nominate their candidates for the general election. The national nominating convention served as the mechanism for coordinating the campaign efforts of the various state and local party organizations that emerged in the 1820s. State and local party leaders dominated the decision because they controlled local caucuses (party meetings) used to select delegates to state conventions that in turn selected delegates to the national conventions. The process of nominating a presidential candidate required building a winning coalition of the party bosses attending the national convention. For more than a century, the national conventions featured political party "bosses" who met during the conventions to negotiate over the nominee, policies, and the spoils of government (i.e., jobs and contracts) in the event of victory. This system of patronage-driven party organizational structure at the national and state levels was conceived of by Martin Van Buren (Andrew Jackson's successor).

The political parties during this time frame existed mainly as coalitions of state and local party organizations. Political power at the time was decentralized among autonomous state and local political parties. This made sense because most government activity at the time occurred at the state and local level where party officials controlled the resources of government. The national government was limited mainly to issues like defense, trade, and allocating land on the frontier. Winning the presidency gave state and local party bosses a source of patronage jobs and contracts that they could use to reward party workers.

The caucus-convention system also worked because campaigns were conducted by local party organizations to get out the popular vote for the party's candidates. Presidential candidates could not campaign for the office directly because of limited transportation and communication. With party organizations controlling nominations and the

21

primary means for campaigning, candidates were dependent on the organizations to get elected. In this kind of system, elected officials were expected to provide patronage jobs and contracts to the party organizations. The system was problematic because representation of the general public interest was often secondary to the interests of the parties. Allegations of corruption were widespread. Governments were often unresponsive to new constituencies like immigrants and people working in the growing manufacturing sector.

The Third Nomination System: The Mixed Caucus-Primary Convention System

Dissatisfaction with government that often protected established interests rather than listen to the demands of citizens led to a series of third party movements in the late 1800s, but these efforts to dislodge the major parties generally failed. By the beginning of the 1900s, however, a new "progressive" reform movement sought to challenge the lock that local party bosses had on government. The progressive movement differed from earlier attempts at reform in that progressive leaders sought to gain access through the major political parties rather than through independent political parties.

Progressive reformers in both political parties had been largely excluded from government by local and state party bosses who controlled nominations and therefore determined who could be elected by voters. These progressives sought to use government to protect citizens from some of the more extreme abuses that occurred in an unregulated capitalist system, advocating policies like a 40-hour work week, child labor laws, and consumer protection laws. The need to gain access to power—to control the authority of government—was a prerequisite to obtaining policy goals. Among other election reforms, progressive politicians promoted primaries as a means to nominate candidates. Primary elections are open to the voters who meet criteria defined by the political parties or the legislatures in each state. Most states restricted participation in primary elections to people who registered to vote and declared their preference for a political party, though some states allowed any registered voter to participate as long as those voters only cast ballots in the primary of one party. By expanding participation in the selection of candidates, primaries expanded the variety of constituencies and demands that elected officials would be beholden to—essentially diluting the power of party bosses and giving progressive politicians a chance to get on the ballot.

While primary elections had been around since the 1880s for nominations to state-level offices, it was not until 1912 that a few states began holding primary elections in presidential nominations. The impact of the reforms was rather limited at the presidential level. Most of these presidential primaries were advisory rather than binding, which meant that the state conventions could follow or ignore the primary vote when selecting delegates to the national conventions. These primaries often were referred to as "beauty contests" since they could establish a candidate's appeal with party voters but did not necessarily result in delegates to the national nominating conventions. Further, some of the states that adopted presidential primaries in 1912 and 1916 reverted to the old system after 1920. The selection of convention delegates continued to be controlled by state and local party bosses for another sixty years. Presidential candidates continued to need the organizational support of state and local party officials who controlled state voting blocs at the national conventions.[4] Senator Estes Kefauver, for example, won 12 of 15 Democratic primaries in 1952, but party bosses rallied around Senator Adlai Stevenson, who was less threatening to their patronage organizations.

None-the-less, over time the results of these nonbinding primaries began to influence the decisions of party bosses, who needed to nominate candidates with popular appeal in order to win the general election.[5] Part of the reason is that, in many parts of the country, progressive era reforms weakened local and state party organizations—especially outside of urban areas and Southern states. One reform—civil service laws that established criteria for holding government positions—removed many of the patronage jobs that provided large numbers of campaign workers for party organizations. Party organizations also lost the ability to monitor how people voted with the widespread adoption of the secret ballot. During the 1800s, the parties often used ballots on different colored paper so they could identify which citizens voted for their candidates. The secret ballot allowed people to vote for candidates with less fear of retribution from party bosses. The expansion of government services and programs in the 1930s also reduced some of the dependency that citizens had on local party organizations. Throughout much of the country, local party organizations were much weaker by the end of World War II. As a result, the party organizations needed to nominate popular candidates if they were going to win elections since they could no longer control voters as they had previously.

Moreover, candidates increasingly used presidential primaries to demonstrate their popular appeal among party constituencies. For example, supporters of Dwight D. Eisenhower used presidential primaries to promote Eisenhower for president in 1952.[6] Eisenhower's popular appeal made him a strategic choice for Republican Party bosses even though he was less conservative than most Republican Party elites. The nomination of Eisenhower illustrates the point that party insiders generally consider winning an election more important than getting a candidate who promotes party ideology. The nomination also illustrates the principle that it is the out-of-power politicians that embrace reforms and technological innovations. In 1952, the Democratic Party had many more party supporters and thus had less incentive to select candidates based on popularity. The Democratic Party ignored the results of the primaries and nominated Adlai Stevenson as their presidential candidate in that same year.

The innovation and proliferation of radio and television also enabled candidates to appeal directly to voters, whereas previously they had to communicate with voters through party organizations. Presidential aspirants increasingly sought the nomination using primaries to demonstrate their popular support. Although John F. Kennedy used the traditional strategy of cultivating party bosses, he also ran in the West Virginia presidential primary to demonstrate that a Catholic from a Northern state could win popular Democratic support in a Southern state.[7] Kennedy cultivated support among local party bosses in West Virginia and appealed to party identifiers with advertising on television. His victory in a Protestant state helped convince party leaders elsewhere in the country that a Catholic candidate could win the general election.

The weakening of local and state party organizations meant that party bosses had less control of nominations even in states holding traditional party caucuses. In 1964, supporters of Barry Goldwater won the nomination in large part by seizing control of the party caucuses in various states. Goldwater's campaign organized political party activists to win delegates in local caucuses to send Goldwater delegates to the state conventions and then to the national convention.[8] Goldwater's candidacy also is notable because it marked the beginning of a transformation in the electoral base of the Republican Party, with Southern white voters starting to identify with the Republican Party.

The important point is that power over nominations during this period gradually shifted from party elites to party activists who had intense demands for policy.[9] The rise of political party activists, whose jobs did not depend on party bosses, has transformed the form and operation of political party organizations. Political parties transitioned from patronage organizations of the 1800s to the more ideological parties that we have today.[10]

The struggle over presidential nominations reached a pivotal moment in 1968 when an opponent of the Vietnam War, Senator Eugene McCarthy, entered the New Hampshire primary to challenge President Lyndon B. Johnson's renomination as the Democratic candidate. Although Johnson won 49% of the vote in that primary, McCarthy's 42% of the vote demonstrated that there was widespread dissatisfaction with the president. Moreover, McCarthy won more delegates to the convention than did Johnson since the allocation of delegates was not tied to the primary vote before 1972. After the New Hampshire primary, a leading anti-war Democrat, Senator Robert Kennedy, entered the race. The outcome of the New Hampshire primary and Kennedy's entry into the race contributed to Johnson declaring that he would not seek nor accept the 1968 Democratic presidential nomination.[11] Anti-war candidates won most of the presidential primaries, but Vice President Hubert Humphrey was nominated at the Democratic Convention without entering a single primary. He won the nomination by appealing to state and local party leaders in states still selecting convention delegates in caucuses and state conventions. Humphrey turned out to be the last presidential candidate selected by appealing directly to party leaders.

The 1968 Democratic Convention in Chicago was highly divisive and violent protests marred the image of the Democratic Party. National television news covered riots on the streets and protests on the convention floor. Liberal anti-war activists were highly dissatisfied with Humphrey's nomination. Many Southern Democrats split from the party in the general election to support either Alabama Governor George Wallace—who ran as an independent candidate—or the Republican candidate Richard Nixon. Humphrey won less than 43% of the national vote and Nixon won the 1968 election.

The Modern Primary and Caucus System

After the disastrous 1968 presidential convention and defeat in the general election, the Democratic Party initiated the "McGovern-Fraser Committee" in 1970 to reform the nomination process.[12] The idea was to come up with reforms that would give greater legitimacy to the party's presidential nominee.[13] The reforms were intended to make the process more open and participatory and to nominate candidates who were more representative of party constituencies. These reforms changed the rules governing who participates in the selection of the president, greatly expanding participation in presidential nominations. Importantly, these changes reflected the reality of a new kind of political party system in which political activists play a much greater role in determining the direction of the parties. Democratic Party insiders agreed to the reforms because they realized that they could no longer rely on party loyalties to win the election.

Although the McGovern-Fraser reforms did not require states to hold primaries, most state legislatures adopted presidential primary elections with binding results that tied the allocation of convention delegates to candidates' shares of the vote in these elections. The number of states holding primaries increased from 15 in 1968 to 21 in 1972 to 27 in 1976 (see Figure 2.1). The majority of delegates to the national conventions have been selected in presidential primaries since 1972. There has been an average of 36 states holding presidential primaries since 1980. There is some variation across elections since some states cancel presidential primaries when only one candidate qualifies for the

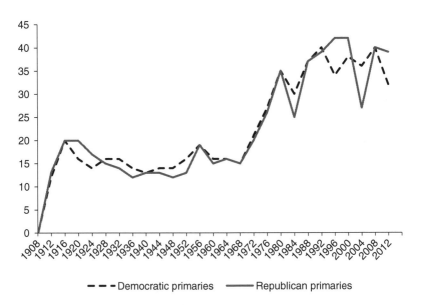

– – – Democratic primaries   ▬▬▬ Republican primaries

Figure 2.1 Number of States Holding Presidential Primaries, 1908 to 2012

Source: Scammon and Scammon, 1994, *Congressional Quarterly Guide to Elections*, Washington DC: Congressional Quarterly Press; www.fec.gov/pubrec/fe1996/presprim.htm; www.thegreenpapers. com.

ballot (which happens when an incumbent president seeks renomination). This expansion of presidential primaries means that candidates need to win the support of large numbers of voters in primaries spread across the country in order to win the nomination.

Those states that retained caucuses and state conventions modified the rules for participation in the caucuses so that these local party meetings today are the functional equivalents of primary elections.[14] A modern-day caucus is a local meeting open to voters similar to what happens in primary elections. Voters meet to discuss candidates and issues and then cast votes for candidates. While there are some differences in the operation of caucuses across states, the number of votes received by each candidate determines how many delegates they will get at a state convention, which in turn selects delegates to the national convention.

The two kinds of nomination elections do require different kinds of political campaigns. Primaries require campaigns of mass appeal with a potentially greater influence for money, advertising, and news coverage. These things are also important to compete for votes in caucuses, but caucuses require more grassroots organization to mobilize and coordinate voters in the local meetings where participants have an opportunity to persuade their friends and neighbors to support particular candidates. The need for grassroots campaigns in caucus states means that organized interest groups have more potential to influence the results of nominations from party caucuses.

After the 1972 election, the Democratic Party effectively gave candidates control over who served as delegates on their behalf at the convention.[15] This reform further reduced the role of state and local party officials at the national conventions. Since these changes were instituted through state legislatures, the Republican Party's presidential nomination process changed as well. The national conventions have been transformed from forums for negotiations over policy and patronage to what they are today—events used by the political parties to showcase their candidate. Since 1972, the results of the presidential caucuses and primaries determine which candidate will become the nominee of his or her political party.

There are some differences in the nomination rules of the two political parties, but the basic structure of binding primaries and caucuses is the same for both. Although there has been some variation in the specific details, after the 1972 and 1976 elections, the Democratic Party adopted rules for allocating delegates to the convention in proportion to candidates' vote shares in caucuses or primaries.[16] Democratic candidates gain delegates to the national convention in proportion to their vote share as long as they get at least 15% of the primary vote. While going along with the move to binding primaries and caucuses, the Republican Party resisted the proportional method of awarding delegates. Most state Republican parties employ winner-take-all, winner-take-most, or modified winner-take-all rules (in which state-wide and/or district level delegates are allocated to the candidate with the most votes). Republicans in different states use different qualifying thresholds in states that mandate proportional delegate allocation, ranging as high as 20% of the vote in order to qualify for delegates to the national convention. There is debate about how much these different delegate-allocation rules affect the nomination campaign. Some argue that winner-take-all rules advantage the front-runner and help unify the party, while others argue that winner-take-all rules have little impact on the outcome of the nomination.[17] Differences in delegate-allocation rules do not appear to make a big difference in how long candidates continue to contest the nomination during the primary season, nor does the difference affect who becomes the nominee.[18] Aside from these differences in delegate-allocation rules, the caucuses and primaries of the two parties operate similarly.

The reforms of the 1970s were the culmination of the transformation of presidential nominations that had been building since the 1950s. The expansion of binding primaries and caucuses democratized the presidential nomination process in the sense that a larger number of voters have the opportunity to participate in the selection of presidential nominees.[19] The reforms shifted influence over the nomination from party bosses and insiders to party activists, campaign contributors, and voters in the reformed primaries and caucuses.[20] The rise of party activists as players in nominations has had a huge impact on national politics. The reforms can be thought of as establishing rules for a different kind of political party in which policy activists play a vital role. Party insiders generally seek candidates who can win elections, often selecting moderate candidates who have appeal to more centrist voters. Party activists, in contrast, feel strongly about certain issues and they demand certain policies. Party activists and organized groups want a candidate who will champion their policies and they generally support more ideological candidates. The rising influence of party activists in party nominations is usually considered to be the main factor contributing to more polarized political parties over the last forty years.

The reforms also opened possibilities for candidates who lacked the support of the traditional party establishment, but who could draw support from issue activists.[21] It is unlikely, for example, that candidates like George McGovern (1972) or Jimmy Carter (1976) would have won their nominations without the rule changes. The move to participatory caucuses and primaries with binding results also changed how candidates compete for a presidential nomination. Before the reforms, candidates needed to secure the support of party leaders at the state and local levels. While winning primaries was helpful for demonstrating electoral appeal, party leaders controlled the nomination. After the reforms, candidates needed to use campaigns of mass appeal to communicate with large numbers of party activists and rank-and-file party identifiers who would vote in caucuses and primaries. These changes increased the importance of money because campaigns of mass appeal are far more expensive. The changes also increased the importance of news media since candidates needed to communicate with large numbers of prospective primary voters. Candidates' campaign organizations and professional consultants also grew more consequential since the campaigns are built around the candidate rather than the party during the nomination campaign.

Some of this was going on previously, of course, since the decline of the local party organizations throughout much of the country gave candidates the need to build their own campaign organizations.[22] To be competitive in the modern era, candidates must build an extensive organization at the national, state, and local levels to identify, communicate with, and mobilize large numbers of potential primary and caucus voters. These organizations have to be built well before the caucuses and primaries begin. In the 2012 Republican nomination campaign, for example, several candidates including Newt Gingrich failed to get on the ballot for the critical Virginia primary because they lacked the organizational ability to get enough signatures to qualify.

Candidates hire professional staffs at both the state and national levels and recruit activists for the grassroots organizational effort. Developments in computer and communications technologies expanded candidates' ability to campaign outside of party organizational networks. Direct mail and social media, for example, enable candidates to reach potential supporters who are not plugged into party networks. This is critical because it increases the potential for a candidate to bring in new voters in caucuses and primaries, which can potentially change the outcome of the nomination race. The media's role as an intermediary between the candidates and the voters also became more important with the increased number of states selecting convention delegates through primaries.[23] Paid advertising and news coverage enable the candidates—who can gain such exposure—to compete for the nomination without extensive support from state and local party officials. The shift to primaries and reformed caucuses also altered campaign strategies. George McGovern—who helped write the new rules better—used the reformed primaries and a few caucuses to gain the 1972 Democratic nomination even though most observers realized that he was too liberal to win the general election. During the 1970s, the expanded primaries enabled relatively obscure candidates to emerge victorious. Most famously, Jimmy Carter used a victory in the Iowa caucus to propel his candidacy from relative obscurity to the Democratic nomination in 1976. Since then, most candidates have copied Carter, concentrating their efforts on the early nominating elections in hopes of gaining momentum. Dark-horse candidates—those who trail in polls or lack money—build their campaign strategy around trying to gain momentum in the early contests in order to grow their share of the vote in subsequent caucuses and primaries.

The competitiveness of the nomination campaigns of the 1970s, however, subsided during the 1980s in part as a result of a number of counter-reforms. These counter-reforms were motivated by a variety of interests and goals, but had the effect of reining in the open quality of the nomination campaigns of the 1970s. First, recognizing that the early states received a lot more attention from candidates and the media, an increasing number of states tried to move the date of their caucus or primary closer to the beginning of the primary season. The purpose was to both allow their voters a meaningful choice among candidates and increase the influence of their state in selecting the party's presidential nominee.[24] Whereas only five states held primaries by the end of March in 1976, about two-thirds of the convention delegates were selected by the end of March by the 1990s.

There is a great deal of debate over which states hold their nominating election first.[25] Largely by chance, Iowa and New Hampshire came to be the first states to hold their nominating elections. Voters in these states have a disproportionate influence on the nomination campaign because candidates who do not finish in the top two vote-getters in these states have a greatly reduced chance of winning (see Table 2.1). These states do not necessarily pick the winner, but they certainly winnow the field of candidates. This is controversial because voters in Iowa and New Hampshire are not representative of the country as a whole. Both are small, rural states that are overwhelmingly white. Given the small unrepresentative character of the states, many argue that these states should not have such a big influence on the nomination. Despite

Table 2.1 Top Candidates Receiving Votes in Presidential Primaries in Open Nominations, 1972 to 2012

Year	Democratic Candidate	Share of the primary vote	Year	Republican Candidate	Share of the primary vote
1972	Humphrey	25.8	1980	*Reagan*	60.8
	McGovern	25.3		G.H.W. Bush	23.3
	Wallace	23.5		Anderson	12.4
	Muskie	11.5			
			1988	*G.H.W. Bush*	67.9
1976	*Carter*	38.8		Dole	19.2
	Brown	15.3			
	Wallace	12.4	1996	*Dole*	59.21
	Udall	10		Buchanan	21.23
				Forbes	10.02
1984	*Mondale*	37.8			
	Hart	36.1	2000	*G.W. Bush*	63.21
	Jackson	18.2		McCain	29.83
1988	*Dukakis*	42.8	2008	*McCain*	46.65
	Jackson	29.1		Romney	22.16
	Gore	13.7		Huckabee	20.12
1992	*B. Clinton*	51.79	2012	*Romney*	52.52
	Brown	20.12		Santorum	20.25
	Tsongas	18.06		Gingrich	14.28
				Paul	10.68
2000	*Gore*	75.66			
	Bradley	19.92			
2004	*Kerry*	60.78			
	Edwards	19.38			
2008	H. Clinton	48.04			
	Obama	47.31			

Note: Includes only candidates receiving at least 10% of the vote across all of the primaries of a political party in a given election year. Votes from caucuses are excluded. Candidates who won the nomination are in italics.

efforts to change the sequence of the caucuses and primaries, however, both states have been able to maintain their privileged position.[26] There are, however, a couple of major advantages to this sequence. One, the small size of the states means that a larger number of candidates can compete for votes. Campaigns in larger states require more money to run advertising. Starting in small states gives candidates who don't have as much money a better chance to appeal to voters and win the nomination, as Jimmy Carter did in 1976. Two, the voters in these states have much more opportunity to make informed decisions about who they will support. Most candidates spend weeks if not months traversing these states, meeting with voters, talking with local reporters, and advertising continuously on radio and TV. There are other states in which caucus and primary voters have as much information about the candidates as do the voters in these states. With respect to democracy, it is preferable that voters make informed decisions. Voters in Iowa and New Hampshire are probably better informed about the candidates compared to those in any other state. The role and influence of these nominating elections will be discussed more in later chapters.

Another counter-reform resulted when some states began to hold primaries on the same date to increase the impact of a region on the nomination. In 1988, nine Southern states created a regional primary that would benefit a more ideologically moderate candidate. Since then, more states have joined in "Super Tuesday" primary dates with a majority of states having primaries or caucuses on six Tuesdays following the New Hampshire primary. The combination of these changes has compressed the primary season into a shorter time frame with little time between primaries for raising money, organizing, or campaigning.

Multi-state "super" primaries advantage presidential candidates with the financial and organizational resources to campaign in several states simultaneously. Candidates with fewer resources are forced to choose where to compete, skipping some states in favor of others. The 2012 Republican nomination campaign illustrates the effects. Front-runner Mitt Romney competed in all nine of the Super Tuesday primaries, while his opponents had to focus on one or two states in an attempt to remain viable. Romney won the most votes in six states, Rick Santorum won two states, and Newt Gingrich won his home state of Georgia. Romney gained 53% of the delegates at stake while Gingrich, Santorum, and Ron Paul divided the remainder among themselves. Romney grew his delegate count and effectively put the nomination out of reach for his rivals.

Another important counter-reform at the national level was the Democratic Party's creation of automatic "super-delegates" for elected officials like governors, U.S. senators and representatives, and national and state party leaders.[27] Super-delegates are not elected by caucus or primary voters, but these officials are able to vote at the national conventions because of their positions as party leaders or elected officials. The number of super-delegates expanded from 1984 until 1992 when all Democratic governors, senators, representatives, big city mayors, and others were given super-delegate status. By 2008, super-delegates accounted for almost a fifth of the delegates at the Democratic Convention. Republicans also have super-delegates, but limit them to members of the Republican National Committee and to each state party chair. Super-delegates cast only about 5% of the votes at the Republican Convention. Super-delegates generally follow the vote in their state caucus or primary and usually do not affect the outcome of the nomination campaign. Super-delegates, for example, sometimes switch their support if their endorsed candidate fails to gain a lot of votes in their state or district.[28]

There are two exceptions to this. In 1984 super-delegates gave Walter Mondale a majority of delegates at the convention (he had about 48% of the delegates gained through caucuses and primaries). Super-delegates also added to Barack Obama's delegate count in 2008. Although Hillary narrowly won more votes in Democratic primaries in 2008, Obama won much more of the vote in states holding caucuses. Along with super-delegates, Obama gained a slight majority of delegates to the convention. Super-delegates give party insiders a notable role in the selection of presidential nominees, something that can be critical in a close race like the 2008 Democratic nomination. As we will see later, party insiders and groups also can impact the nomination above and beyond using their formal voice at the convention.

The contemporary presidential nomination system features candidates competing for the support of party stakeholders and party identifiers who vote in caucuses and primaries. The reforms of the early 1970s codified party activists' participation in the process. Party activists are more intense in their political and policy preferences than the average American. The rise of these policy demanders changed the politics of presidential nomination campaigns, so that candidates seeking the nomination need to gain the support of these more ideological liberals (in the Democratic Party) and conservatives (in the Republican Party). The expansion of the voice of party activists has increased the polarization of political parties in the United States.

There is some tension and struggle between party activists and party insiders.[29] Activists push for candidates who are deeply committed to policy positions that are perceived by non-partisans as relatively extreme. There is some evidence that Republican Party activists currently place more rigid demands on their elected officials in Congress than do Democrats. For example, a Pew Center poll in October 2013 found that higher percentages of Republicans—especially those Republicans identifying as members of the Tea Party movement—expected their elected officials to follow the preferences of their district even if the elected official believed a different action was in the best interest of the country.[30] Democrats, in comparison, are more evenly split on this question. The greater intensity of demands being made of Republican office holders has contributed to the asymmetric polarization of the political parties since the 1990s. The Republican Party has drifted to a more conservative position and the Democrats have moved in a liberal direction. The threat of a primary challenge by Tea Party–backed candidates has limited the discretion of Republican legislators to negotiate and compromise in office. While more conservative candidates may have greater appeal to Republican activists, these more ideologically extreme candidates have been less successful appealing to voters in the more diverse populations in U.S. Senate elections or in presidential primaries. Most party insiders recognize this and sometimes push back against party activists.

The implications for democracy and representation are mixed. On one hand, expanding participation makes the nomination process more democratic and representative of party constituencies. On the other hand, the parties' nominees may be more ideologically extreme relative to the policy preferences of the general public. This potentially makes the choices offered in the general election less representative of the preferences of the general population. The irony is that efforts to make party nominations more democratic may lessen the extent to which general elections produce a president who represents the broad interests of the country as a whole. Citizens voting in the general election potentially face a choice between more ideologically extreme candidates who are less representative of their preferences.

Notes

1 James W. Ceaser, 1979, *Presidential Selection*, Princeton, NJ: Princeton University Press.
2 Ceaser, 1979, *Presidential Selection*.
3 The Electoral College balances representation of states and population. Each state has a number of Electors to the Electoral College equal to the state's number of Representatives that are apportioned on the basis of population. Obtaining a majority

in the Electoral College requires a combination of a geographic and a popular vote majority.

4   Leon D. Epstein, 1986, *Political Parties in the American Mold*, Madison: University of Wisconsin Press, 89–95.

5   Reiter, 1985, *Selecting the President.*

6   Wayne Steger, 2012, "Dwight D. Eisenhower, 34th President of the United States," in *Chronology of the U.S. Presidency*, Matthew Manweller (ed.), Westport, CT: Greenwood Press, 1065–1116.

7   James W. Davis, 1967, *Presidential Primaries: Road to the White House*, Westport, CT: Greenwood Press.

8   Philip E. Converse, Aage R. Clausen, and Warren E. Miller, 1965, "Electoral Myth and Reality: The 1964 Election," *American Political Science Review*, 59(2): 321–336; Theodore Rosenoff, 2003, *Realignment: The Theory that Changed the Way We Think about American Politics*, New York: Roman and Littlefield, 95–98.

9   Reiter, 1985, *Selecting the President*; Martin P. Wattenberg, 1984, *The Decline of American Political Parties, 1952–1980*, Cambridge, MA: Harvard University Press.

10 Hans Noel, 2013, *Political Ideologies and Political Parties*, New York: Cambridge University Press.

11 Doris Kearns Goodwin, 1991, *Lyndon Johnson and the American Dream*, 8th ed., New York: St. Martins.

12 Austin Ranney, 1975, *Curing the Mischiefs of Faction: Party Reform in America*, Berkeley: University of California Press; Byron E. Shafer, 1983, *Quiet Revolution: Reform Politics in the Democratic Party, 1968–1972*, New York: Russell Sage; William Crotty, 1977, *Political Reform and the American Experiment*, New York: Crowell.

13 Jeanne J. Kirkpatrick, 1978, *Dismantling the Parties: Reflections on Party Reform and Party Decline*, Washington DC: American Enterprise Institute Press.

14 Barbara Norrander, 1993, "Nomination Choices: Caucus and Primary Outcomes, 1976–1988," *American Journal of Political Science*, 37(2): 343–364.

15 David E. Price. 1984, *Bringing Back the Parties*, Washington DC: CQ Press, 214–217.

16 Herbert B. Asher, 1984, *Presidential Elections and American Politics*, 3rd ed., Homewood, IL: Dorsey Press, 199–201.

17 Elaine C. Kamark and Kenneth Goldstein, 1994, "The Rules do Matter: Postreform Presidential Nominating Politics," in *The Parties Respond*, 2nd ed., L. S. Maisel (ed.), Boulder, CO: Westview Press, 165–195.

18 Norrander, 2006, "The Attrition Game"; Brian Arbour, 2009, "Even Closer, Even Longer: What if the 2008 Democratic Primary used Republican Rules?" *Forum*, 7(2).

19 Ceaser, 1979, *Presidential Selection.*

20 Kirkpatrick, 1978, *Dismantling the Parties.*

21 Crotty, 1977, *Political Reform*; Steger, Hickman, and Yohn, 2002, "Candidate Competition and Attrition in Presidential Primaries."

22 Martin P. Wattenberg, 1991, *The Rise of Candidate Centered Politics*, Cambridge, MA: Harvard University Press.

23 Thomas E. Patterson, 1980, *The Mass Media Election*, New York: Praeger.

24 Larry J. Sabato, 1997, "Presidential Nominations: The Front-Loaded Frenzy of '96," in *Toward the Millennium: the Elections of 1996*, Larry J. Sabato (ed.), Boston: Allyn and Bacon, 37–92.

25 William G. Mayer and Andrew E. Busch, 2004, *The Front-loading Problem in Presidential Nominations*, Washington DC: Brookings Institution; Hugh Gregg and Bill Gardner, 2003, *Why New Hampshire: The First-in-the-Nation Primary State*, Nashua,

NH: Resources-NH; Thomas Gingale, 2008, *From the Primaries to the Polls*, Westport, CT: Praeger; Andrew E. Smith and David W. Moore, 2015, *Out of the Gate: the New Hampshire Primary and its Role in the Presidential Nomination Process*, Dartmouth, NH: University Press of New England.

26 Smith and Moore, 2015, *Out of the Gate*.

27 Price, 1984, *Bringing Back the Parties*; Epstein, 1986, *Political Parties in the American Mold*.

28 Kenny J. Whitby, 2014, *Strategic Decision-Making in Presidential Nominations: When and Why Party Elites Decide to Support a Candidate*, Albany: State University Press of New York.

29 Richard L. Butler, 2004, *Claiming the Mantle: How Presidential Nominations Are Won and Lost Before the Votes are Cast*, Boulder, CO: Westview Press.

30 "Tea Party's Image Turns More Negative," Pew Research Center, Washington DC (October 16, 2013), www.people-press.org/files/legacy-pdf/10-16-13%20Tea%20 Party%20Release.pdf.

Part II

MECHANICS OF THE GAME

3

PARTY STAKEHOLDERS—THE INSIDER GAME

The reforms of the early 1970s moved the formal selection of presidential nominees from the national conventions to the caucuses and primaries held in states during the first six months of the election year. The proliferation of primaries and reformed caucuses expanded the ability of party activists and party identifiers to participate in the selection of presidential nominees.[1] In this sense, the reforms democratized the presidential nomination process by putting the formal selection of presidential nominees in the hands of the active portion of the mass memberships of the political parties. The reforms, however, did not remove party stakeholders from the process. Stakeholders try to influence the selection of nominees before the mass membership of the party has a chance to vote.

Presidential nomination campaigns begin much earlier, during what is commonly called the invisible primary—the early phase of the campaign occurring during the year or so prior to the election year when candidates and party stakeholders engage each other in the search for leadership. Prospective candidates reach out to party stakeholders to gauge their potential backing and get pledges of support. Party stakeholders engage in a "long national discussion" among themselves about which candidate they believe will be able to promote party policy positions and who can win the election.[2] They seek to persuade each other as well as the larger number of party activists and party identifiers about which candidate should be nominated.[3] If a critical mass of party insiders, activists, and groups throw their support behind a candidate, that candidate becomes highly likely to gain the nomination during the caucuses and primaries. The invisible primary thus is mainly an insider game largely played by candidates and party stakeholders. The outcome of this insider game can affect the selection of the nominee—even before votes are cast in the caucuses and primaries.

Modern political parties are networks of party leaders, elected officials, activists, and aligned groups who aspire to hold power for the purpose of controlling public policy.[4] It should be noted that these party stakeholders can be divided into two groups with differing priorities that can lead to different preferences for candidates.[5] Group leaders and party activists are policy demanders who want a nominee who will champion their issue or policy cause first and foremost.[6] They want an authentic candidate who can be trusted to follow through on their promises to deliver on policies that are important to these party constituencies. Party leaders and elected officials form a second group of party stakeholders. These party elites have similar policy goals but they differ from party activists and group leaders in that they usually want a candidate who is electable first and foremost. Party elites, however, are also beholden to their own constituents so they tend to support a candidate who will appeal to the party activists in their electoral districts or states.[7] Thus the candidate preferences of most party and elected officials generally are similar to those of party activists in their districts or states. If there is a conflict between them, party elites either refrain from making an endorsement or they switch their support to the candidate who is gaining momentum during the caucuses and primaries.[8] Given the substantial correspondence between the endorsements of party elites on one hand and party activists and group leaders on the other, the endorsements of party elites is a pretty good indicator of party stakeholder support for presidential nomination candidates.

To a significant degree, the invisible primary functions as the equivalent of the national nominating conventions of the pre-reform era. The interactions among party stakeholders participating in the long national discussion function as did the backroom negotiations during conventions in the pre-reform era, although without the deal-making and exchanges of support for patronage. Party stakeholders interact in an attempt to coordinate among themselves to select a presidential nominee that best represents their interests and who can win in the general election.[9] They signal each other through the media and online forums about who they think should be selected. If a critical mass of party insiders, activists, and groups throw their support behind a candidate, that preferred candidate becomes highly likely to gain the nomination during the caucuses and primaries. The preferred candidate gains a substantial lead in endorsements, fundraising, media coverage, and support in national public opinion polls. The front-runner goes on to win the nomination because caucus and primary voters tend to go with the choice of party stakeholders—when stakeholders come to agreement on who should be nominated as the presidential candidate of the party. While there is substantial communication among party networks before the caucuses and primaries begin, party stakeholders usually— but not always—come to agreement on the nominee.[10] If they do reach a consensus, then the candidate who earns the support of the various party constituencies will gain an enormous advantage in the presidential nomination campaign. There have been nominations in which party stakeholders fail to reach a sufficient consensus about a preferred candidate to help that candidate become the nominee.

Table 3.1 shows the candidate with the most endorsements by elite elected officials (presidents, governors, senators, and U.S. Representatives) during the invisible primaries for the election years 1976 to 2012. The endorsements of these officials are the most reliably and most consistently reported in newspapers and magazines. Endorsements by other officials are haphazardly reported by newspapers and other media, making the measurement of such endorsements less reliable. The table shows only open nomination campaigns—those without an incumbent seeking reelection—because incumbent presidents have won renomination even when they have been challenged, for example, President

Table 3.1 The Most-Endorsed Candidate's Share of the Endorsements of Elite Elected Officials by the End of the Invisible Primary, 1976 to 2012

Election year	Most Endorsed Democratic Presidential Candidate	Percent of Elite Endorsements	Most Endorsed Republican Presidential Candidates	Percent of Elite Endorsements
1976	Lloyd Bentsen	34.7	Incumbent reelection	
1980	Incumbent reelection		*Ronald Reagan*	50.9
1984	*Walter Mondale*	57.4	Incumbent reelection	
1988	Richard Gephardt	40.3	*George H. W. Bush*	63.1
1992	*Bill Clinton*	62.0	Incumbent reelection	
1996	Incumbent reelection		*Bob Dole*	65.3
2000	*Al Gore*	93.5	*George W. Bush*	82.6
2004	Howard Dean	25.3	Incumbent reelection	
2008	Hillary Clinton	60.1	Mitt Romney	31.6
2012	Incumbent reelection		*Mitt Romney*	67.7

Note: Figures are the percentage of weighted endorsements received by each candidate. The weights assigned to each endorsement are from Marty Cohen, David Karol, Hans Noel, and John Zaller, 2008, *The Party Decides: Presidential Nominations Before and After Reform*, Chicago: University of Chicago Press. Winning candidates are in italics (for Democrats 3 out of 7, for Republicans 5 out of 6).

Ford in 1976 and President Jimmy Carter in 1980.[11] The proportion of elite endorsements is an indicator of the extent to which party insiders are willing to support a particular presidential candidate. When a candidate gained a majority of these elite endorsements by the end of the invisible primary, that candidate has won the nomination in eight of nine nomination campaigns (2008 is the exception when Senator Barack Obama won despite having about a third as many elite endorsements as Senator Hillary Clinton). The Democratic nomination campaigns of 1984, 1992, 2000, and 2008 and the Republican nomination campaigns of 1980, 1988, 1996, 2000, and 2012 all can be characterized as having party elites achieving substantial elite consensus about which candidate should be nominated. In all but the 2008 Democratic nomination, the preferred candidate of party elites went on to win the nomination. Democratic Party elites failed to reach any kind of agreement about a preferred candidate in 1976, 1988, and 2004. In these elections, one candidate had a plurality (the most but not a majority) of party endorsements and failed to win the nomination. On the Republican side, Mitt Romney had slightly more endorsements from elite elected officials than had John McCain in 2008 but still lost the nomination.

In presidential nomination campaigns in which party elites reach a large degree of agreement during the invisible primary about which candidate should be nominated, that candidate has become the nominee in eight of nine open nominations. Party elites failed to reach a consensus during the invisible primaries in 4 of 13 open nominations since 1976. In these years, the candidate with the most elite endorsements failed to win the nomination of his or her party. It appears that party insiders have substantial influence on the presidential nomination when they come to agreement on which candidate should be nominated, but party insiders have failed to come to agreement in almost one-third of the nomination campaigns. These nominations are up for grabs when the caucuses and primaries begin (see chapter six).

Why Party Stakeholder Endorsements Matter

Party stakeholders signal their support for a candidate by endorsing a candidate and helping that candidate compete for the support of the mass membership of the party. We can track stakeholder support by observing endorsements for the candidates. Endorsements have two kinds of effects on the nomination campaign. The first is informational. The second is an indirect effect on the nomination though the influence on candidates' relative abilities to compete for votes. These two effects are intertwined and occur more or less simultaneously during a campaign. Endorsement information is communicated on a national scale through the news media, blogs, and social media. Attentive publics—activists, contributors, and aligned groups—are more likely to be exposed to this information.[12] These attentive publics provide candidates with the resources needed to compete for the support of large numbers of primary voters across the country. Candidates who receive more endorsements get more attention from the national media. They are able to raise more money. They benefit from signals that tell attentive publics that a candidate is preferable, viable, and electable. As a result, endorsed candidates have more ability to compete for the support of large numbers of caucus and primary voters who may not even be aware of the endorsements, but who will none-the-less be affected in their candidate preferences.

The support of party stakeholders helps a candidate build a campaign capable of appealing to large numbers of party activists and party identifiers. Party stakeholders are a part of candidates' efforts to build their own networks of volunteers and contributors from the overlapping organizations and networks that form the modern political parties. Candidates use endorsements in their fundraising appeals and campaign communications. Party elites and group leaders frequently serve as headliners at fundraising events with or on behalf of presidential candidates. They may even operate as an extension of the endorsed candidate's campaign organization by encouraging members of their own political networks to support the endorsed candidate. Candidates who lack the support of party stakeholders have a harder time attracting campaign contributions and building a campaign organization.

Party elites also influence media coverage of presidential nomination campaigns. In an uncertain environment, journalists, editors, and producers look for cues about what to cover and how to cover it. Journalists pay close attention to polls and quarterly financial reports to figure out which candidates are leading, lagging, rising, and falling. But journalists also look for insider perspectives because objective indicators are sometimes misleading, especially in competitive campaigns with multiple candidates. National news reports frequently incorporate subjective elite judgments that are not reflected in objective indicators of the horse race.[13] Party insiders communicate frames or spin on the campaign, interpreting events and scenarios in ways that help the candidate that they are backing. The value to a candidate is that supportive party stakeholders talk up the preferred candidate to other party insiders, donors, and the media.[14] Party elites also may serve as proxies in attacks on rivals, which are effective because party and elected officials are "credible" sources of criticism and may insulate the candidate from charges of negative campaigning.[15] Since elite support for—and criticism of—candidates is often covered in the news, party elites contribute to the perceptions conveyed to the public about the candidates.

The support or opposition of party stakeholders provides information about candidates to the mass membership of the party during a nomination campaign. Most voters in a general election simply vote for the candidate of the political party that they identify with. During a nomination campaign, however, voters cannot use party labels because all of the candidates are from the same party.[16] Political elites play a greater role in the formation of public opinion when public awareness is low.[17] Endorsements and party insider commentary provide party members with information about the candidates' ideological orientations and policy positions.[18] In a field of candidates saying similar things, commentary and endorsements by party elites help voters distinguish between candidates. This information is particularly important early in a campaign when large numbers of prospective voters are unfamiliar with the candidates. The nomination campaign is an uncertain environment in which candidate support is soft and can shift as party activists and party identifiers learn about the candidates.[19] In such an environment, indicators of elite support like endorsements provide cues about which candidates are personally and ideological acceptable and which candidates can win.

Party stakeholders' support for candidates helps prospective caucus and primary voters in another way. Indications of party stakeholder support tell party activists and party identifiers which candidates are worth paying attention to and which candidates can or should be ignored. Presidential nomination campaigns usually have a number of candidates who are vying for caucus and primary voters' attention and support. These voters do not evaluate all of the candidates and pick the best candidate. Rather, most voters in the earliest nomination elections limit their choices to the nationally credible candidates.[20] They reduce the costs of becoming informed by focusing on a few candidates and eliminating candidates from serious consideration on the basis of awareness and viability.[21] This process gives party stakeholders the opportunity to influence the candidate preferences of caucus and primary voters by sending cues about which candidates are preferable, viable, and electable. Caucus and primary voters' preferences for candidates tend to follow party elite endorsements in part because these tend to be the candidates that voters think about and consider. Candidates lacking party elite support are often ignored because voters don't think they can win.

Endorsements convey meaningful information because party stakeholders pay attention to candidates' personal characteristics and policy positions as well as evidence of candidate chances of success.[22] As such, endorsements convey perceptual information about which candidates are personally and ideologically preferable. Party stakeholders want an ideologically compatible candidate that they can work with and they do not endorse candidates who are unacceptable on policy grounds. Elite party office-holders also are motivated to have the best candidate who can win. They want a candidate at the top of the ticket who will help or at least not hurt candidates like themselves who appear lower on the ballot during the general election.[23]

Still, the impact of endorsements is a matter of dispute. Newspaper articles often quote campaign and party officials to the effect that endorsements do not carry much weight with most primary voters.[24] These claims are easy to discount because the claims are usually made by candidates who lack endorsements. Still, it is the case that sometimes nominations are won by candidates other than the one with the most endorsements (see Table 3.1). For example, Barack Obama beat Hillary Clinton for the 2008 Democratic nomination even though she had more endorsements.

Detractors also argue that citizens pay too little attention to be aware of endorsements. The least attentive citizens, however, are unlikely to vote in presidential caucuses and primaries. The party activists and party identifiers who participate in caucuses and primaries are somewhat more attentive and are exposed to more information about the nomination campaigns from the news media and campaign advertisements. Elites influence the formation of public opinion most among citizens who pay sporadic attention to politics and who exhibit moderate levels of political awareness.[25] A study by the Annenberg Public Policy Center shows that most voters in the early caucuses and primaries were exposed to at least some information about endorsements during the 2000 presidential nomination campaigns.[26] The study found, for example, that 78% of the voters in the New Hampshire Democratic primary knew that Senator Edward Kennedy had endorsed Al Gore. The study found that endorsements did influence some votes in these elections.

It is generally the case that an individual endorsement matters less than the overall pattern of endorsements. One reason is that the impact of an individual endorsement depends on whether a citizen is aware of it and whether they agree or disagree with the person or group making the endorsement. Citizens generally are not well informed about events. They get bits and pieces of information that they obtain through episodic attention to politics. If voters don't know about an endorsement, then that endorsement cannot influence their thinking about the candidates. An individual endorsement is more likely to be missed by a given voter than a series of endorsements since there is an increased chance that the voter will become aware of an endorsement.

The effect of an individual endorsement also depends on the audience's impression of the endorsing person or organization. People interpret information in ways that are consistent with their existing beliefs, especially their party loyalties, ideological orientations, and political attitudes.[27] A person is more likely to view an endorsement favorably if the person has a favorable view of the endorsing organization or politician; otherwise the endorsement may have the opposite effect. An endorsement by Senator Ted Cruz, for example, may help or hurt a Republican presidential candidate depending on a voter's opinion of Cruz. People who like Senator Cruz may be more likely to have a positive view of the endorsed candidate while people who think Cruz is too extreme may be less likely to support the endorsed candidate. Thus, an individual endorsement may have a limited impact on the nomination campaign.

While an individual endorsement may have a limited effect on a caucus or primary voter's preference for a presidential candidate, the overall pattern of endorsements is important for understanding the impact of insider support for a candidate. First, a large number of endorsements also increase the chances that a given voter will become aware of at least some of these endorsements. Second, information becomes more potent as patterns emerge and become discernible. The overall pattern of endorsements enables people who are paying attention to discern the ideological bases of candidates' support and to make judgments about candidates' viability and electability. Candidates who gain large numbers of endorsements are signaled to be ideologically acceptable, viable, and electable because they are receiving support from across the spectrum of the party's elite membership. People tend to accept opinion leadership from credible sources, but an individual endorser may not be a credible source for a given person. The odds improve that a given party identifier will find a credible endorser as more party officials endorse a candidate. Having a lot of endorsements tells us that a lot of stakeholders think a candidate would be a good choice. If, for example, a Republican candidate is endorsed by conservative Republican Senators like Ted Cruz and James Inhofe *and* by moderate Republicans like

Senator Mark Kirk and Susan Collins, then Republican voters know that the endorsed candidate is acceptable to both moderates and conservatives. The candidate benefits because more of the Republican voters are inclined to perceive the candidate as acceptable and maybe even preferable to other candidates seeking the nomination.

Candidates who gain endorsements from a small number of stakeholders are signaled to have limited appeal—usually defined by geography or ideological allies. In the 1996 Republican nomination race, for example, Senator Richard Lugar received endorsements only from elected officials in his home state—indicating that he had limited support beyond his friends in Indiana. Most of the candidates who seek a presidential nomination gain only a few endorsements from elite office-holders, which is a clear signal that these candidates lack widespread appeal among party insiders. The absence of elite endorsements may signal that a candidate lacks viability or has unacceptable personal characteristics or policy positions. For example, Republican Representative Jack Kemp ran for the 1988 Republican nomination, but failed to get any endorsements from elite Republicans of his home state. The absence of support from the politicians who presumably knew Kemp well was taken by many as an indication that there was a problem with his candidacy. Thus the overall pattern of endorsements tells us which candidate has acceptable policy positions, is viable, and may be electable—which can influence the media, party officials, activists, contributors, and aligned groups who are attentive to the race.

Endorsements also affect voters in *later* primaries indirectly through their impact on voters in *early* caucuses and primaries. Voters in the Iowa caucus and the New Hampshire primary are exposed to much more attention to the nomination campaign—these voters are bombarded with television advertising and the local news have extensive coverage of the candidates.[28] Voters in other states get less information about the candidates. The voters in Iowa and New Hampshire matter because endorsements have the greatest effect in the early caucus and primary states in which candidates invest heavily in time and money.[29] The results of early caucuses and primaries in turn affect the vote in later states.[30] Thus endorsement information may affect voters in later states indirectly by affecting the vote in Iowa and New Hampshire.

Early Endorsements Matter the Most

Endorsements should have an impact on the nomination campaign when party stakeholders coalesce and endorse candidates early—when the media, activists, and contributors need information about who they should pay attention to. When more party stakeholders support a candidate, they collectively signal that the candidate is more desirable and viable. A consensus among the various party establishments reflects a widely held belief that a particular candidate is preferable from a personal and policy standpoint, is viable, and is electable. The less consensus among stakeholders about who should be supported and the later in time that endorsements occur, the weaker the signal the party stakeholders send to each other and to their attentive partisan publics. A candidate who receives a disproportionate share of endorsements probably has the characteristics that make him or her likely to win. That candidate likely will appeal to political party constituencies across the country. That candidate will receive more media coverage and will have help in raising money. These factors combine to make it likely that a candidate with a disproportionate share of party insider support will be able to gain the support of party activists and identifiers before the caucuses and primaries begin.

The more insiders and activists divide their support among the candidates or refrain from making endorsements, the greater the likelihood that the candidates are not very appealing to the full range of the party constituencies and the less likely that any one candidate will emerge as a strong front-runner during the invisible primary. In this scenario, the nomination race will be more competitive as multiple candidates enter the caucuses and primaries with a decent chance of winning the nomination.

Party stakeholders may refrain from public support of a candidate for various reasons. Some Democrats have refrained from making an endorsement as a result of being designated as super-delegates—a status created by Democrats in 1982 for high profile party and elected officials. At least through 1992, the Democratic National Committee discouraged super-delegates from making an endorsement before the caucus and primary voters had a chance to weigh in. Others have refrained from endorsing candidates because of personal relationships with several of the candidates. For example, some Republican Senators in 1988 refrained from any endorsement to avoid alienating either Senate Minority Leader Bob Dole or Vice President George H. W. Bush when both sought the nomination. Others may avoid making an endorsement because they don't like the choices. White Southern Democratic office holders have been less likely to endorse a presidential candidate before the caucuses or primaries. These office holders sometimes prefer to avoid associating with a "liberal" Democrat. Finally, party insiders may not know which candidate is going to emerge as the candidate with the most appeal with party constituencies. Uncertainty deters endorsements because party insiders are wary of endorsing a candidate that their own constituents may end up rejecting. Whatever the reason, a lack of early commitments of support is a pretty good sign that the nomination race is open and competitive. The lack of commitments by party insiders contributes to and reinforces this competitiveness by denying the media, party activists, and donors a signal about which candidate is preferable, viable, and electable. If stakeholders remain divided or undecided about which candidate should be the nominee, then the critical period of selection will necessarily move on to the caucuses and primaries. Endorsements occurring after the caucuses and primaries begin generally matter less than those occurring before that time.[31] Once party activists and identifiers start voting in the Iowa caucus and the New Hampshire primary, most of the attention shifts to how well candidates are doing in those contests. Far less attention is given to party insiders who make endorsements after that time. These endorsements are usually considered to be jumping on the bandwagon of surging candidates rather than as attempts to influence the vote in subsequent caucuses and primaries. So the window of opportunity for party stakeholders to influence the selection of presidential nominees is mainly during the invisible primary. If stakeholders unify behind a candidate during that time, then that candidate is going to have a strong probability of winning the nomination. If stakeholders fail to unify sufficiently during the invisible primary, then it is the decisions of voters in the early caucuses and primaries that will have relatively more influence on the selection of the nominee.

Thus party stakeholders can influence the selection of presidential nominees by unifying more behind one candidate than others. The candidate with the most endorsements during the invisible primary will be more able to raise money, build a stronger campaign organization, and will receive more news media coverage conveying that they are the preferred candidate in terms of policy, viability, and electability. If enough party stakeholders back a candidate during the invisible primary, that candidate becomes likely to get the support of a majority of voters in the caucuses and primaries that select delegates to the national nominating conventions. If, however, elites remain divided or undecided, then there is a less clear signal to campaign donors, the media, or to party activists and identifiers as to which candidate should be the nominee. In this case, the nomination remains open and competitive going into the caucuses and primaries where party voters will have the determinative voice in selecting the nominee.

In terms of the implications for democracy, there is a difference between nominations in which party insiders rally behind a candidate during the invisible primary and those nominations in which party insiders remain divided and undecided. Nominations in which party insiders reach a sufficient degree of unity on a candidate can be said to be mediated by party stakeholders. Party stakeholders essentially collude to determine the outcome by stacking the deck in favor of their preferred candidate. Nominations in which party stakeholders fail to reach an agreement among themselves (and they do), result in the voters in presidential primaries and caucuses having more options. These nominations can be said to be more democratic in that more citizens are able to express a meaningful voice in the selection of the party nominees.

Notes

1 Theodore Lowi, 1979, *The End of Liberalism*, New York: W.W. Norton.
2 Cohen et al., 2008, *The Party Decides*.
3 Cohen et al., 2008, *The Party Decides*.
4 Cohen et al., 2008, *The Party Decides;* Bawn et al., 2012, "A Theory of Parties"; Seth Masket, 2009, *No Middle Ground: How Informal Party Organizations Control Nominations and Polarize Legislatures*, Ann Arbor: University of Michigan Press.
5 Butler, 2004, *Claiming the Mantle*.
6 Cohen et al., 2008, *The Party Decides*; Bawn et al., 2012, "A Theory of Parties."
7 Whitby, 2014, *Strategic Decision-Making in Presidential Nominations*.
8 Steger, 2008, "Inter-Party Differences in Elite Support for Presidential Nomination Candidates"; Whitby, 2014, *Strategic Decision-Making in Presidential Nominations*.
9 Cohen et al., 2008, *The Party Decides*.
10 Steger, 2013, "Two Paradigms of Presidential Nominations."
11 Wayne P. Steger, 2003, "Presidential Renomination Challenges in the 20th Century," *Presidential Studies Quarterly*, 33(4): 827–852.
12 Wayne P. Steger and Christine Williams, 2011, "Analysis of Social Network and Traditional Political Participation in the 2008 Elections," Paper presented at the Annual Meeting of the American Political Science Association, Seattle, WA.
13 Chris Cillizza, 2006, "The Friday Line: Winning the 2008 Money Primary," *Washington Post*, February 3.
14 Whitby, 2014, *Strategic Decision-Making in Presidential Nominations*, 26.
15 Gina Garramone, 1985, "Effects of Negative Political Advertising: the Roles of Sponsor and Rebuttal," *Journal of Broadcasting and Electronic Media*, 29: 147–159.
16 Mark J. Wattier, 1983, "Ideological Voting in 1980 Republican Presidential Primaries," *Journal of Politics*, 45(4): 1016–1026.
17 John R. Zaller, 1992, *The Nature and Origins of Mass Opinion*, New York: Cambridge University Press; Scott Keeter and Cliff Zukin, 1983, *Uninformed Choice: The Failure of the New Presidential Nomination System*, New York: Praeger; Patrick J. Kenney, 1993,

"An Examination of How Voters Form Impressions of Candidates' Issue Positions During the Nomination Campaign," *Political Behavior*, 315(2): 265–288.

18 Jamieson, Johnston, and Hagen, 2000, "The 2000 Nominating Campaign"; Gregory Neddenriep and Anthony J. Nownes, 2012, "An Experimental Investigation of the Effects of Interest-Group Endorsements on Poorly Aligned Partisans in the 2008 Presidential Election," *Party Politics*, http://ppq.sagepub.com/content/early/2012/06/15/1354068811436067.

19 Samuel Popkin, 1991, *The Reasoning Voter*, Chicago: University of Chicago Press; Rebecca Morton and Kenneth Williams, 2000, *Learning by Voting: Sequential Voting in Presidential Primaries and Other Elections*, Ann Arbor: University of Michigan Press.

20 Christopher C. Hull, 2008, *Grassroots Rules: How the Iowa Caucus Helps Elect American Presidents*, Stanford, CA: Stanford University Press.

21 Walter J. Stone, Ronald Rapoport, and Lonna Rae Atkeson, 1995, "A Simulation Model of Presidential Nomination Choice," *American Journal of Political Science*, 39(1): 135–161.

22 Richard L. Berke, 1996, "Variety of Motives in the Making of Endorsements," *New York Times*, March 14.

23 Butler, 2004, *Claiming the Mantel*.

24 See, for example, David S. Broder, 1999, "Showy Bandwagon Is No Free Ride to a Bush Nomination," *Washington Post*, March 8.

25 Zaller, 1992, *The Nature and Origins of Mass Opinion*.

26 Jamieson, Johnston, and Hagen, 2000, "The 2000 Nominating Campaign."

27 G. R. Boynton and Milton Lodge, 1994, "Voters' Images of Candidates," in *Presidential Campaigns and American Self Images*, Arthur H. Miller and Bruce E. Gronbeck (eds.), Boulder, CO: Westview Press, 176–189.

28 Smith and Moore, 2015, *Out of the Gate*.

29 Hull, 2008, *Grassroots Rules*; Jamieson, Johnston, and Hagen, 2000, "The 2000 Nominating Campaign."

30 Wayne P. Steger, Randall E. Adkins, and Andrew J. Dowdle, 2004, "The New Hampshire Effect in Presidential Nominations," *Political Research Quarterly*, 57(3): 375–390.

31 Steger, 2013, "Two Paradigms of Presidential Nominations."

4

THE CHANGING MONEY GAME

Presidential nominations began changing in the 1950s and 1960s, becoming more "candidate centered." Candidate-centered campaigns focus on the individual candidate who aspires to become the face and image of their political party. Conducting such a candidate-centered campaign became much more expensive after the reforms of the 1970s, which forced candidates to seek the support of millions of caucus and primary voters.[1] The high costs of presidential nomination campaigns advantage candidates who can raise money. It also provides an opportunity for party stakeholders to tilt the playing field in favor of their preferred candidate. Coordination among party stakeholders can help a candidate gain a competitive advantage in the campaign by helping that candidate raise more money. A candidate with more money can run a more professional campaign and advertise to reach large numbers of voters.

Campaign fundraising occurs in a political marketplace. In order to raise funds, candidates need to have the personal characteristics and policy positions that will appeal to party activists who are the main source of funds in political campaigns. In a sense, candidates have to have the right stuff—what it takes to win—in order to raise the money that will enable them to win. Long-shot candidates usually cannot raise much money until they demonstrate that kind of appeal.[2] Money can be a mechanism that separates winners from losers in a campaign. No candidate—regardless of their ideas or qualifications—can win without money. But only those candidates—who have what it takes to attract support—can raise a lot of money. The effect is self-reinforcing—candidates who have widespread appeal can raise the money needed to expand their support; candidates who don't catch fire with party activists are unable to do the things they need to attract support and they necessarily stay mired in the back of the pack.

When Campaign Funds Matter—Competitive and Non-Competitive Races

Money matters because modern campaigns cost a lot. A competitive presidential campaign will be run by professionals with expertise at developing strategy, fundraising, organizing, conducting research, developing messages and advertisements, and coordinating grassroots volunteers. Campaigns also require a lot of money for communicating with voters through multi-media campaigns. Money gives a candidate the capacity to communicate to prospective voters why he or she should be elected and not some other candidate. Campaign funds are a critical determinant of how competitive an election will be. If one candidate has a lot more money than other candidates, that candidate has a greater ability to explain to voters why they should support him or her. Candidates with less money have less ability to promote themselves or to explain to voters why another candidate would be a worse choice.

Candidates must have money but money does not determine the winner. Having money is a necessary but not sufficient condition to win a presidential nomination campaign. Money enables a candidate to make their pitch to voters; it does not mean that voters will buy it. The history of political campaigns is filled with examples of candidates who had a lot of money and failed to win the election. The effect of money on a campaign is asymmetric. Having a large campaign war chest does not ensure victory, but a lack of money ensures defeat. Without money, no candidate can win a presidential nomination regardless of their policy ideas or personal charisma. What money does do is give a candidate a better *chance* to attract supporters than a candidate who lacks campaign funds.

Thus money makes a difference in a campaign as the imbalance of funds gets large—when one candidate has a lot more money than other candidates. If campaign funds become concentrated in the hands of one candidate—because most donors give their money to the same candidate—then that candidate gains substantial advantage in the campaign. Voters will know more about the well-funded candidate than they will know about less well-funded candidates. The candidate with more money will have more ability to present favorable information about him or herself and to criticize other candidates. Poorly funded candidates will not have as much ability to tell voters about themselves or the flaws of the well-funded candidate. In this scenario, voters will learn mostly favorable things about the well-funded candidate and it becomes less likely that they would vote for someone else.

53

If candidates have the same amount of campaign funds, then they would have about the same capacity to make their case to voters and none would have a particular advantage over the others. Candidates would have similar organizations, similar amounts of advertising, and similar ability to get their supporters to the polls in a caucus or primary. This kind of competitive race is preferable from the standpoint of voters. A competitive race provides voters with more information about all of the candidates, which enables them to be more informed about their choices.[3] Further, competitive campaigns usually feature better quality information. This empowers voters because they have more information about the candidates when they make their decision about who they will support. In this respect, campaigns in which multiple candidates have roughly similar amounts of money can be said to be more democratic than those in which candidates have unequal amounts of campaign funds.

This makes it possible for donors—mainly party activists who care about policy—to have a big impact on the race.[4] Party activists and members of groups aligned with the political parties are the main source of campaign contributions. How much money a candidate can raise is an indication of the candidate's appeal among party activists and groups aligned with the party. Candidates who have a lot of appeal among party activists tend to raise more money. Because party activists and groups are the main source of campaign funds, it becomes possible for party stakeholders to collude to advantage one candidate over others. If party activists and groups come to an agreement on which candidate to support, then they can enable their chosen candidate to run a more comprehensive campaign with greater ability to persuade voters. As party stakeholders concentrate their contributions, they effectively narrow the pool of candidates that caucus and primary voters will choose among—making the caucuses and primaries less democratic than they appear to be. If campaign contributors divide their support among multiple candidates, then the candidates have greater parity in their resources and thus in their ability to make their case to voters.

While campaign fundraising gives party stakeholders the opportunity to collude by concentrating their donations to a preferred candidate, the distribution of campaign funds is much less concentrated than are the endorsements of party insiders, as discussed in chapter three. Looking at campaign contributions, there usually is no clear favorite to whom party donors have given their money. Instead, campaign contributions usually are spread among the candidates in presidential nominations—though campaign funds generally are not *evenly* distributed across the candidates. Table 4.1 shows the share of funds raised

Table 4.1 The Leading Fundraiser's Share of Money by the End of the Invisible Primary, 1976 to 2012

Election year	Leading Fundraiser Among Democratic Candidates	Percent of Funds Raised	Leading Fundraiser Among Republican Candidates	Percent of Funds Raised
1976	George Wallace	24.7	Incumbent reelection	
1980	Incumbent reelection		*Ronald Reagan*	30.9
1984	*Walter Mondale*	40.9	Incumbent reelection	
1988	*Michael Dukakis*	28.8	Pat Robertson	28.8
1992	*Bill Clinton*	33.9	Incumbent reelection	
1996	Incumbent reelection		*Bob Dole*	26.4
2000	*Al Gore*	51.0	*George W. Bush*	42.6
2004	Howard Dean	30.1	Incumbent reelection	
2008	Hillary Clinton	36.7	Mitt Romney	34.0
2012	Incumbent reelection		*Mitt Romney*	38.0

Source: Federal Election Commission reports for the last time period prior to the Iowa caucuses (FEC Form 3P, page 2, line 22).

Note: Winning candidates are in italics.

by the candidate raising the most money in the invisible primaries from 1976 to 2012 in open presidential nominations. In only one open presidential nomination race did one candidate raise more than half of the funds raised by all of candidates combined. This is what would be expected if donors coordinated their efforts to finance a particular candidate. That candidate was Vice President Al Gore who raised 51% of the money raised by a Democratic presidential candidate in 1999. Gore's only opponent, however, raised 49% of the funds, so these candidates were almost evenly matched in their pre-primary fundraising. In most races, there are two or three candidates who raise a substantial amount of money. While there is not always parity in their campaign funds, these candidates have enough money to make the race competitive at least through the earliest caucuses and primaries.

The candidate who raised the most money before the primaries won the nomination about two-thirds of the time between 1976 and 2012. The candidate who raised the most money during the invisible primary won four of seven open Democratic nominations and four of six open Republican nominations. George Wallace raised more money than any other Democratic candidate in 1976, but he was beaten by Jimmy Carter who had raised less than half as much money during the invisible primary. The circumstances of Carter's campaign have not been repeated—the subsequent front-loading of the primary schedule has limited the amount of time that candidates have to raise money after the initial nominating elections. Candidates have needed to finish near the top of the money chase to have a good chance of winning the nomination. But raising the most money during the invisible primary does not ensure victory, as evidenced by George Wallace, Pat Robertson, Howard Dean, Hillary Clinton, and Mitt Romney (in 2008). Raising money during the invisible primary probably has become less consequential by the 2000s than it was in the 1980s and 1990s. The emergence of the internet now enables candidates to raise money very quickly. In 2004, Howard Dean lost the race despite raising the most money in the invisible primary. In 2008, Hillary Clinton raised the most money by the end of 2007, but she was defeated by Barack Obama, who had raised only half as much as Clinton by that time. Mitt Romney raised slightly more than John McCain ahead of the 2008 Republican primaries, but McCain won the race. Money enables candidates to compete but it is not sufficient to win. Understanding how money has influenced campaigns in different ways at different points of time requires looking at how the financing of campaigns has changed over the past fifty years.

The Continuously Changing System of Campaign Finance

The financing of presidential nomination campaigns continuously changes as a result of federal campaign laws and candidates adapting to changing technology. Federal campaign finance regulations affect how much money candidates can raise and spend. Historically, presidential candidates relied on a few big donors to finance their campaigns. Campaign finance regulations enacted in the 1970s limited the sums that candidates could raise from individual donors. Recent developments in campaign finance law, however, have enabled big donors to once again weigh in on presidential nomination campaigns with huge contributions. Innovations in technology matter because candidates tend to be quick to adapt new technologies to give themselves an advantage in the competition for the nomination. The internet is potentially a game changer by enabling candidates with popular appeal among party activists to raise large sums of money from millions of donors—potentially neutralizing the influence of billionaire donors. These developments have made presidential nominations more competitive and less predictable from the standpoint of campaign funds.

The rules governing campaign finance affects candidates' ability to raise money. Prior to the 1970s, presidential nominations and elections were funded mainly by very wealthy donors. Congress enacted the Federal Election Campaign Act (FECA) of 1971 and substantially amended the law in 1974 to limit the influence of wealthy donors in federal elections. The Federal Election Commission (FEC) was created to enforce the rules and to administer a public funding program for presidential campaigns. The FECA was intended to limit the influence of wealthy donors by imposing limits on how much money candidates could accept from an individual donor, limiting how much candidates could spend in each state, and requiring them to disclose who gave them money and how they spent it. In addition, the law tried to level the playing field among presidential candidates by providing *matching funds* to candidates. The federal government would match the first $250 of every contribution raised by candidates, provided the candidate could qualify by raising money from individuals in at least 20 states. Matching funds helped level the playing field by giving dark-horse candidates the money they need to conduct a presidential nomination campaign. Candidates who accepted matching funds had to adhere to limits on spending in states, with the cap being determined by the population of each state. States with smaller populations like Iowa and New Hampshire had low spending caps, which enabled a larger number of candidates to compete on a relatively even footing in these elections.

Campaign finance regulations of the 1970s also limited how much money candidates could accept from individual donors and from political action committees. Candidates were limited to accepting donations of up to $1,000 for individuals and $5,000 for political action committees. Political action committees associated with interest groups generally give little money to presidential campaigns, so most of the money has to be raised from individuals. Thus political activists are the main source of funds in a presidential campaign. Note that candidates are regulated. FEC does not regulate how much money individuals can spend or how much a candidate can spend on their own campaign. The FECA regulations originally limited how much money an individual could spend, but the Supreme Court struck down limits on how much an individual could spend independently of a candidate's campaign organization in the case of *Buckley v. Valeo* (1976). After that case, wealthy donors could contribute $1,000 to a candidate—the maximum a candidate could accept—and they could spend as much as they wanted beyond that as long as they didn't coordinate their spending with the campaign (more on this point later).

Candidates for federal office, including the presidency, are not limited in how much of their own money they spend to get elected. Several multi-millionaire candidates have spent large sums of money seeking a presidential nomination. Among the lesser known include Morry Taylor, who spent about $6 million seeking the 1996 Republican nomination. Despite spending a lot of money, Taylor's campaign was virtually unknown beyond Iowa and New Hampshire. More prominently was Steve Forbes' campaign for the 1996 Republican nomination. Despite spending more than $37 million of his own money, Forbes won only the Arizona and Delaware Republican primaries.[5] These cases illustrate a couple of important points. One, spending money does not equate to votes. Two, the money spent by candidates who self-finance is not a good indicator of their political appeal. Candidates who raise money from others have demonstrated that they have enough appeal among party activists that these activists are willing to give to the candidate. Self-financed candidates can make their case to voters, but they have not had much success in presidential nomination campaigns. Forbes' campaign, however, had the effect of changing campaign finance forever. Fearing a well-funding challenge from Steve Forbes in the 2000 Republican nomination, George W. Bush decided to refuse matching funds.[6]

From 1976 through the 1990s, campaign finance regulations forced candidates to raise funds from a large number of donors. For example, if a candidate wanted to raise $50 million dollars, the minimum number of individual donors would have been 50,000 citizens donating the maximum of $1,000 per person. Raising money from large numbers of people requires a huge fundraising operation on a national scale. That kind of system advantaged well-known candidates who had developed a network of supporters across the country. Before the internet, candidates raised money through networks of donors and through direct mail solicitations of money from party activists. Candidates had to start running for the presidential nominations very early because they needed time to build networks of campaign contributors on a national scale.[7] Candidates who had national name recognition and who were tied into national networks had an advantage in fundraising during this time frame.

The proliferation of computers greatly facilitated direct mail solicitations because organizations could track donations and develop lists of "proven" donors. In the 1970s, this strategy was new. Professional consultant Richard Viguerie began the direct mail boom with lists of voters on main frame computers. The use of direct mail mushroomed even more with the low cost of personal computers. Initially, professional fundraisers didn't know who would or would not donate, so they spent large sums of money mailing large numbers of potential donors—few of whom actually donated money. In 1976, for example, George Wallace raised more money than any of the other Democratic candidates, but he spent nearly all of the money he raised on direct mail used to raise the money. As a result, Wallace had little left over to use for his campaign. The costs of direct mail declined in the 1980s as professional fundraisers became better at targeting donors, resulting in more money raised per dollar spent on direct mail. This produced more money available for the campaigns during the 1980s and 1990s. One of the biggest challenges for potential candidates considering a run for the presidency has been to secure the services of professional fundraisers. A candidate who could get the best fundraisers gained a substantial advantage in the money chase.

Initially, changes in campaign finance regulations made presidential nomination campaigns more competitive. The campaign finance reforms encouraged "outsider" and "ideological" candidates to run.[8] The 1976 FECA provided "matching funds" to candidates. Matching funds helped candidates who receive larger numbers of small contributions rather than the maximum contributions. State-by-state spending limits also limited the advantages of the better-financed candidates, which leveled the playing field in the early caucus and primary states. Lesser-known candidates could concentrate resources in early states like Iowa and New Hampshire in the hope of gaining a break-through victory that would create media exposure, increase name recognition, and generate fundraising for later primaries. While candidates figured out ways to get around these restrictions, the overall effect was to level the playing field in the earliest caucuses and primaries.

The system of federal matching funding and spending caps contributed to dark-horse candidates adopting the same "break-through" strategy in which they focus nearly all their efforts in Iowa and New Hampshire. If they beat expectations in these states' nominating elections, then they could gain momentum and continue in the race. Candidates who received more votes in these states' nominating elections generally gained a boost in their fundraising efforts. Jimmy Carter epitomized the "momentum" campaign. Carter began the 1976 primary season with relatively little money, but he was able to capitalize on his stronger than expected share of the vote in Iowa and New Hampshire. He had a month to raise money before the next primaries and he raised enough money to continue competing for the nomination. If dark-horse candidates fail to do well in Iowa or New Hampshire, they are hard-pressed to raise additional money. Further, candidates who fail to get at least 10% of the vote in two consecutive primaries are cut off from matching funds—effectively shutting down their campaigns. In order to become eligible again for matching funds, they must gain 20% of the vote in a primary. In general, candidates who raise a lot of money during the invisible primary—but who do not spend a lot of it—may be able to overcome doing poorly in Iowa or New Hampshire because they have the financial resources to compete in subsequent primaries. Candidates who have to spend everything they have on these elections are generally forced out of the race.

Changing technology and fundraising strategies have rendered obsolete the 1970s campaign finance rules. In the 2000 election cycle, George W. Bush decided to forego federal matching funds. This helped him because he could ignore the state-by-state limits on spending, which meant that he could raise and spend far more than other candidates. George W. Bush fundraised over $70 million by January 31, 2000, while his closest competitor John McCain had just over $21 million. At the beginning of the caucuses and primaries, Bush had over $20 million in cash reserves while the McCain campaign had about $350,000. Though McCain won the New Hampshire primary and became a media sensation, he could not compete with Bush's well-financed campaign in subsequent caucuses and primaries.[9] Since 2000, all of the major candidates in both parties have declined federal matching funds so they can be free to spend as much as they can raise.

Campaign regulations were also amended by the Bipartisan Campaign Reform Act of 2002 (BCRA). The BCRA intended to limit the impact of "soft money" in federal campaigns—particularly congressional races, in which the political party organizations were raising and spending huge sums of money. The law limited funds raised and spent by political party organizations on federal campaigns. The law also sought to limit "issue advocacy ads," which are essentially political ads attacking or supporting a political candidate (though technically not advocating the election of a particular candidate). BCRA banned these ads within 30 days of a primary or within 60 days of the general election. This section of the law was struck down by the Supreme Court as an unconstitutional restriction on freedom of speech.

The BCRA doubled to $2,000 the maximum amount that a candidate could accept from an individual donor for a nomination campaign (and another $2,000 for the general election). This higher limit also is tied to the rate of inflation so it increases each election cycle. This provision enabled candidates to raise more money from big donors. For example, 68% of the $715,150,163 raised by Barack Obama in 2012 was from individual donors giving more than $200.[10] Mitt Romney raised $443,363,010, of which 82% came from large individual donors.[11] Being able to raise more money from big donors increases the incentive for candidates with fundraising ability to ignore the spending limits associated with federal matching funds.

The BCRA also altered campaign finance by spurring the rise of 527s—groups that effectively circumvented the BCRA's soft money ban. A 527 is a group registered under Section 527 of the Internal Revenue Code. These groups are required to file disclosure reports to the FEC if they engage in certain kinds of electioneering communications. These groups initially appeared to play little role in the presidential nominations and were involved mainly in the general election campaign in which there are much greater ideological differences between Democrats and Republicans. According to the Center for Responsive Politics, 527 groups spent over $561 million in the 2012 election cycle, of which a little over $155 million was targeted to federal elections.[12] The vast majority of that amount was spent on congressional elections rather than the presidential election. Thus 527 groups have not had much effect on presidential nominations.

A far more significant change in the system of campaign finance regulation occurred in the Supreme Court case *Citizens United v. Federal Election Commission* (2010). As a result of this decision, corporations, unions, and issue advocacy organizations gained the ability to spend unlimited amounts of money from their treasuries on independent political expenditures in support of or opposition to a candidate. That means companies or unions could spend money on television ads or other forms of advertising expressly advocating for the election or defeat of a candidate. The Supreme Court ruling is often misinterpreted as allowing companies or unions to donate to candidates but that interpretation was explicitly rejected in the ruling. Critics of the ruling are concerned mainly about the huge financial power of corporations to essentially buy the office, essentially undermining the ideal of "one man–one vote." Critics of the ruling recognize that business interests (corporations and investors) have vastly greater resources than do labor groups and individual citizens. Concerns about corporate America directly buying elections, however, may be somewhat overblown. Few corporations, to date, have spent much directly advocating for or against candidates in federal elections. Rather these groups may be directing their money to a new kind of political organization.

The precedent of the 527 groups combined with the *Citizens United* decision contributed to the emergence of organizations called 501(c) groups, which are registered under section 501(c) of the Internal Revenue Code. There are a variety of 501(c) groups that engage in a range of activities for charitable, religious, educational, social welfare, agricultural, labor, or business purposes. Groups organized for social welfare purposes under section 501(c)(4) of the Internal Revenue Code can engage in political activities as long it is not their main purpose. This is a murky area and these groups have engaged in a variety of political activities with limited disclosure of their donors. The *Citizens United* decision by the Supreme Court enables corporations to give funds from their general treasury to 501(c)(4) groups without threat of bad publicity or consumer backlash. The threat of unleashed corporations heavily favors Republican candidates. Between 2010 and the end of the 2012 election cycle, conservative groups qualifying as 501(c)(4) status—for social welfare purposes—outspent liberal groups by a margin of 34 to 1.[13] This margin did not count spending of several groups, including Crossroads GPS, a group affiliated with Karl Rove which spent heavily in support of Republican candidates and in opposition to Democratic candidates. This "anonymous" source of money has much greater potential impact on presidential nominations.

The unlimited sums of money that these groups can spend on a race have changed the relationship between campaign finance and nomination outcomes. Traditionally, candidates needed to raise large sums of money from numerous individuals in networks, which took time and organization. Candidates who raised a large sum of money could develop strategies to outlast their opponents. The emergence of these independent groups, however, means that money can come into the nomination campaign at almost any time, anywhere, and in any amount in support of or in opposition to a candidate.

The emergence of 501(c) groups added uncertainty to the calculations and strategies of the candidates, and makes nomination races potentially more volatile during the caucuses and primaries (which we will discuss later). Candidates who may not be able to raise money from large numbers of donors but who have the support of billionaires might be more willing to run in presidential nominations, and this can change the strategies of all the other candidates in the race.

It should be noted that candidates bankrolled by billionaires are often weaker candidates than the size of their campaign war-chests would suggest. These candidates certainly can staff a campaign organization with professional consultants and put out slick advertising. Their money, however, does not mean that they have widespread support. Candidates who raise money from large numbers of donors are stronger candidates because they have demonstrated that they have widespread appeal among party activists. Candidates who have money, but who have not demonstrated this kind of appeal among party activists, can run and scare front-runners but they may or may not be able to win a presidential nomination (more on this later). Outside groups are not the only ones playing this game. Recall that the federal campaign finance regulations limit what a candidate can accept from a donor. Those donors are free to donate as much as they want to "independent" groups that support the same candidate. Candidates set up their own political action committees (PACs) to promote their party and their own campaigns. Initially these PACs were called leadership PACs. Ronald Reagan pioneered this aspect of campaigning by establishing a "leadership" PAC which the candidate uses to raise money to distribute to other candidates running for elective office. A leadership PAC enables a presidential candidate to curry favor with party stakeholders in hopes of getting the favor returned when they run for president. People who are considering a run for the presidential nomination of their political party raise money for other candidates, in an expectation of support when they do run. Note that these leadership PACs don't spend much directly on the presidential campaign, but are a political party building effort done to gain support for a candidate by people in the party.

Leadership PACs have been supplanted by "Super PACs" that raise huge sums of money that is spent directly on the presidential candidate's campaign. Mitt Romney's Super PAC, Restore Our Future, spent over $142 million in the 2012 election cycle.[14] Of this, 62% was spent against Barack Obama but almost 28% was spent attacking Romney's rivals for the 2012 Republican nomination. One of the big advantages of being an incumbent president seeking reelection is that the incumbent rarely has to fight opponents within his own party for the nomination. For example, Barack Obama's Super PAC, Priorities U.S.A., spent a little over $65 million during the race—all attacking Romney. The emergence of Super PACs has effectively wiped out the limits set by campaign finance regulation. Campaign finance in presidential nominating campaigns operates according to a political market subject to the laws of supply and demand in which candidates with the most appeal among party activists (or billionaires) are the ones who can raise the most money.

Because fundraising operates as a political market, the concentration of money in the campaign tells us about the preferences of party

activists who are willing and able to donate. Candidates who raise a
lot of money from large numbers of donors have demonstrated that
they are appealing to a large number of party activists who in turn
show that they are sufficiently enthusiastic about the candidate that
they are willing to give money to help that candidate win. Candidates
who raise money from a small number of big donors generally don't
demonstrate that they have the same kind of support. Candidates who
receive the largesse of billionaire donors, however, do gain an ability to
communicate with party activists and identifiers in ways that they otherwise would not be able to afford. These donors thus add uncertainty
to a nomination campaign because these candidates may be able to persuade party activists and identifiers that they are worth a second look.
As important, these kinds of billionaire-financed candidates can attack
the front-runner—essentially telling party activists and party identifiers
why the front-runner is not a good candidate.

The 2012 Republican nomination featured a handful of billionaires who preferred more conservative candidates and they potentially
changed the race by contributing huge sums of money to those candidates through super PACs.[15] Sheldon Adelson, a billionaire from the
casino industry, contributed $20.5 million to one of Newt Gingrich's
five Super PACs. Texas billionaire Harold Simmons gave Gingrich
another $1.1 million. Together, these two billionaires financed $21.6 million of the $24 million raised and spent by Gingrich's Winning Our
Future Super PAC. Without these funds, Gingrich would not have been
taken seriously in the Republican primaries. Almost $13 million of that
money was spent supporting Gingrich while most of the rest was spent
attacking Romney. Only $5,000 was spent attacking Barack Obama.
Similarly, Foster Fries, a conservative Christian, and William Dore, an
investor in the energy sector, contributed over $5.5 million to Republi-
can Rick Santorum's Super PAC, the "Red, White, & Blue Fund."[16] The
ability and willingness of ideologically motivated billionaires to finance
candidates has made the nomination campaign more competitive and
more uncertain.

Since billionaires can give massive sums of money in a short period
of time, the old strategy of spending years to build national networks
of donors no longer eliminates candidates who cannot build such networks. In 2012, for example, a handful of billionaires helped make
candidates like Gingrich or Santorum—who lacked widespread support among party insiders or party identifiers—into competitive candidates. Candidates like these used to have short-lived campaigns because
they couldn't compete beyond the earliest caucuses and primaries. As

recently as 2000, George W. Bush could eliminate most of his rivals by beating them in early contests. Long-shot candidates who failed to do well in these early elections simply could not raise the additional funds needed to continue competing with the front-runner in subsequent nominating elections. The possibility of billionaires weighing in on the race by donating huge sums of money means that even dark-horse candidates could get the resources needed to compete in primaries. Even the possibility of an unexpected surge of money forces front-runners to change their strategy, raise even more money, and build a campaign infrastructure to withstand a well-funded opponent. Nomination campaigns have become less predictable as a result.

The internet also has destabilized campaign finance in presidential nominations—but in a way that works in the opposite direction of billionaire donors. There are two aspects to how the internet has impacted presidential nomination campaigns. First, it is possible for candidates to raise large sums of money in small donations from party activists in a short period of time. According to an analysis by the Campaign Finance Institute and the Institute for Politics, Democracy & the Internet, a total of 625,000 small donors gave money to a major-party presidential candidate in 2000. In the 2004 race, that number surged to between 2 million and 2.8 million.[17] The Obama campaign of 2008, for example, developed a strategy of asking millions of small donors (people giving $5, $10, or $20) to donate repeatedly across the months of a campaign. Collectively, the sums of money are substantial. Candidates are no longer dependent on a small number of bundlers and professional fundraisers who have access to networks of donors. Candidates can use the internet to raise funds quickly and cheaply. The internet was critical in the fundraising efforts of John McCain in 2000, John Kerry in 2004, and Barack Obama in 2008. Each of these candidates received more votes than expected in Iowa or New Hampshire. Each of them gained a huge surge in their fundraising, which enabled them to compete in subsequent caucuses and primaries. The internet has revitalized campaign momentum as a potential game changer in presidential nominations because candidates can raise large sums of money in a short time span.

During the 1980s and 1990s, the front-loaded caucus and primary season favored the early front-runner because that candidate had the money, media attention, and organization to compete simultaneously in numerous primaries. Dark-horse candidates, lacking name recognition and other indicators of a viable campaign, desperately needed to "break through" with a victory in the earliest caucuses or primaries.

Front-loading, however, limited the opportunities of these dark-horse candidates for capitalizing on momentum because they have little time to convert momentum into usable resources. While Jimmy Carter had a month to raise money and build an organization after his break-through victory in the 1976 Iowa caucus, candidates since then have had about a week before the next primaries.

The internet changed the effective time needed to raise money and to communicate with prospective voters who become interested in a candidate.[18] John McCain's campaign, for example, reported raising $2.2 million dollars after the 2000 New Hampshire primary.[19] While that was insufficient up against the financial juggernaut of the George W. Bush campaign, candidates in recent nomination campaigns have translated early victories into big campaign donations. Most notably, Barack Obama's initial presidential campaign in 2007 to 2008 relied heavily on the internet for communicating with Democratic Party activists and donors. Obama's 2008 online fundraising operation received 6.5 million donations from more than 3 million distinct contributors who gave an average donation of $80, adding up to more than $500 million.[20] Obama's 2012 fundraising operation topped $1 billion during the entire campaign—a sum that was made possible largely through millions of donations solicited through email, social media, and the campaign website.

Advances in campaign targeting have vastly expanded the ability of a presidential campaign to target potential donors as well. Anticipating a tough reelection campaign, the Obama campaign spent four years building a more refined and integrated database for use in micro-targeting fundraising appeals, campaign messaging, and Get-Out-The-Vote (GOTV) operations.[21] Throughout the nomination and general election phases of 2012, the Obama campaign raised funds through digital means including a Quick Donate program that allowed people to donate repeatedly via text message. When counting only fundraising that was initially generated by digital efforts, including email, social media, mobile phone apps, and the website, the 2012 Obama campaign raised $504 million, up from $403 million in 2008.[22] The Romney and Republican National Committee digital fundraising effort raised $182 million through online contributions.[23] Clearly, the internet has become a major factor in fundraising. It also serves as a counterbalance to the influence of billionaires and donor networks built by party insiders. The vast majority of the contributions going to nomination campaigns through digital methods were in contributions of less than $200. These are the donations of millions of political activists who want their candidate to win.

The internet does not change the basic relationship between candidate-appeal and fundraising, but it does have the potential to influence candidate and party activist decisions in ways that could change the outcome of a presidential nomination campaign. Candidates still must be appealing to party activists in order to raise money from them. Candidates who fail to raise money typically lack the personal or policy appeal that grabs the attention of party activists and excites them enough to donate. What has changed is that donors do not have to be plugged into an existing network of donors to be solicited for a contribution. A party activist who is excited by a candidate can go to the candidates' site and make a donation. In a sense, the internet has vastly expanded the participatory opportunities for millions of people who previously would not have been contributors to presidential nomination campaigns. This makes a presidential run tempting for a politician who lacks a national network but who believes that he or she will be appealing to party activists and thus can raise the money needed to be a credible candidate for the race. The very possibility of raising money this way could change the calculus of candidates considering a run for the presidential nomination of their political party. The implications of this are the subject of chapter nine.

Overall, the financing of presidential nomination campaigns has undergone a mixture of changes that have made them both more and less democratic. Supreme Court rulings have unleashed the potential of a handful of billionaire donors to give tens of millions of dollars to candidates. This development undermines the principle of "one person—one vote" that is a hallmark of democratic elections. On the other hand, the internet opens the possibilities to allow millions of average Americans to get involved in the selection of the nation's leaders by donating $5 or $10 or $20. This effectively expands the donor franchise to a larger number of citizens. In both cases, money has become just as much an element of citizen participation in elections as voting has been historically. It is no longer enough to focus on how many people vote to evaluate the quality of democracy in America. We also need to look at how many people are willing and able to contribute to candidates. The poorest citizens are left out of this equation, losing voice in the selection of presidents.

It is worth noting that poor and lower middle income people have never had much voice in political campaigns. Money has always been a factor in political campaigns, but lower income people have not been able to contribute. The über wealthy have always been able to contribute to campaigns. The political class has and continues to interact with the wealthiest Americans, so the recent removal of limits on campaign contributions has not greatly changed the influence of the wealthiest citizens. Indeed, the era of limited campaign contributions in place from 1976 to the early 2000s is a historical anomaly that resulted from Democratic Congresses seeking to limit the influence of the rich on elections. The development of the internet, however, is a potential game changer by enabling millions of middle-class Americans to give small donations to politicians. The wealthiest citizens have regained their financial voice in campaigns for federal office. The middle class have gained for the first time the opportunity to express their voice through campaign contributions.

Notes

1 Aldrich, 1980, *Before the Conventions*; Wattenberg, 1991, *The Rise of Candidate Centered Politics*.
2 Diana Mutz, 1995, "Effects of Horse Race Coverage on Campaign Coffers: Strategic Contributing in Presidential Primaries," *Journal of Politics*, 5(4): 1015–1042.
3 Keena Lipsitz, 2011, *Competitive Elections and the American Voter*, Philadelphia: University of Pennsylvania Press.
4 Andrew Dowdle, Scott Limbocker, Song Yang, Karen Sebold, and Patrick A. Stewart, 2013, *The Invisible Hands of Political Parties in Presidential Elections: Party Activists and Political Aggregation from 2004 to 2012*, New York: Palgrave Pivot.
5 www.opensecrets.org/news/2011/12/unlimited-presidential-fundraising.
6 www.opensecrets.org/news/2011/12/unlimited-presidential-fundraising.
7 Joel H. Goldstein, 1978, "The Influence of Money in the Pre-Nomination Stage of the Presidential Selection Process: The Case of the 1976 Election," *Presidential Studies Quarterly*, 8(1): 164–179.
8 Nelson W. Polsby, 1983, *Consequences of Party Reform*, New York: Oxford University Press; Wattenberg, 1984, *The Decline of American Political Parties*; Wattenberg, 1991, *The Rise of Candidate Centered Politics*.
9 Randall E. Adkins and Andrew J. Dowdle, 2004, "Bumps on the Road to the White House," *Journal of Political Marketing*, 3(1): 1–27.
10 www.opensecrets.org/pres12/candidate.php?id=N00009638.
11 www.opensecrets.org/pres12/candidate.php?id=N00000286.
12 www.opensecrets.org/527s/index.php#summ.
13 Robert Maquire, 2013, "Conservative Groups Granted Exemption Vastly Outspent Liberal Ones," www.opensecrets.org/news/2013/05/conservative-groups-granted-exemption-vastly-outspent-liberal.html.
14 www.opensecrets.org/pacs/indexpend.php?strID=C00490045&cycle=2012.
15 www.opensecrets.org/pacs/lookup2.php?strID=C00507525&cycle=2012.
16 www.opensecrets.org/pres12/candidate.php?id=N00001380.
17 James A. Barnes, 2008, "Online Fundraising Revolution," *National Journal*, www.nationaljournal.com/magazine/online-fundraising-revolution-20080419.
18 Phillip Paolino and Daron R. Shaw, 2003, "Can the Internet Help Outsider Candidates Win the Presidential Nomination?" *PS: Political Science and Politics*, 36(2): 193–197.

19 Barnes, 2008, "Online Fundraising Revolution."
20 Jose A. Vargas, 2012, "Obama Raised a Half Billion Dollars," http://voices.washington post.com/44/2008/11/obama-raised-half-a-billion-on.html.
21 Wayne P. Steger, 2013, "A Transformational Presidential Campaign: Marketing and Candidate Messaging in 2012," in *Winning the Presidency, 2012*, William Crotty (ed.), Routledge, 74–89.
22 Michael Sherer, 2012, "Obama's 2012 Digital Fundraising Outperformed 2008," *Time*, http://swampland.time.com/2012/11/15/exclusive-obamas-2012-digital-fundraising-outperformed2008/#channel=f18eb1100d5e662&origin=http%3A%2F%23Ffb_xd_fragment%23xd_sig%3Df2c9e929e9d73b2%26, November 15.
23 Zac Moffat, 2012, www.targetedvictory.com/2012/12/success-of-the-romney-republican-digital-efforts-2012/, December 11.

5

MASS- AND MICRO-MEDIA CAMPAIGNS

Candidates in the post-reform era must obtain the support of millions of primary and caucus voters across numerous states. Reaching that many voters through advertising is cost-prohibitive. Even the best funded candidates cannot advertise in all of the states in which they must compete. Candidates depend in large part on the free coverage provided by print and broadcast news media to reach prospective voters. The news media can advance or hinder the cause of particular candidates by giving them more or less exposure, labeling candidates in more or less favorable terms, paying attention to certain issues and not others, and by setting expectations for candidate performance.[1] Candidates also campaign online through low-cost social media, interactive websites, blogs, and YouTube. The internet has changed nomination campaigns, enabling campaigns to engage in micro- or individual-level targeting and messaging in a campaign. How candidates seek the nomination evolves as candidates adapt innovations in communications, conducting mass-media and micro-media campaigns for the support of prospective caucus and primary voters.

News coverage of candidates is critical to the competition among candidates. Candidates need media coverage to build name recognition, communicate ideas, and appeal for support. No candidate can win a modern presidential nomination without the free coverage provided by the news media because caucus and primary voters are not going to support a candidate they do not know. Like campaign finance, news coverage is necessary but not sufficient to win a presidential nomination. Candidates cannot win without getting news coverage, but news coverage does not determine who will win.

Unlike party insiders and campaign contributors, the news media are not in the business of trying to select the candidate who will champion particular policies. The news media—even the more ideological cable news programs—are for-profit businesses. They need viewers and readers to stay in business. That motivation leads them to cover nomination campaigns in ways that will attract a news audience. The amount of coverage of candidates closely follows their standing in the polls. The quality of news coverage—whether favorable or unfavorable to the candidate—also tends to track their standing in the polls. The substantive focus of news coverage tends to be what is understandable and interesting to the audience. While the news media are often criticized in politics, they are providing the content that the public will watch or read.

What works for the news organizations and what attracts the attention of the public, however, is not necessarily what is needed for democracy. Prospective voters need media coverage of the candidates and their ideas if they are to make the kinds of rational vote decisions prescribed in various theories of representative democracy.[2] In an ideal world, the news media engage in watch dog journalism in which they investigate candidates' backgrounds and provide detailed and truthful information about candidates' policy positions. As a practical matter, journalists often report on information provided by political insiders and the campaign operatives of the candidates. Given the time constraints of the news cycle, journalists often have to rely on biased sources that have their own motivations and prejudices. In this sense, it is important to have experienced journalists reporting on the campaigns because these reporters have a better ability to recognize the biases of political sources. The national media tend to have more expertise in campaign reporting compared to the local media, which cover campaigns less frequently. The national media also tend to be more credible sources of information than commentary on the internet, which varies widely in accuracy and reliability. The declining economic viability of national news organizations is a serious concern to the functioning of democratic elections in America because citizens will have to rely on even less reliable and accurate sources for information about candidates.

The competitiveness of the campaign affects the volume and quality of reporting on candidates. Voters get the most detailed information about the backgrounds of candidates when there are several candidates who are competitive in the campaign.[3] Part of the reason is that competitive campaigns attract more interest in the public so news organizations give these campaigns more coverage. Another reason is that competitive campaigns have more information available to reporters. In campaigns with one candidate dominating the race, there tends to be less information about the candidates, and voters will make their decisions based on less-than-complete information. Journalists have multiple sources of information and angles to report on when there are several candidates who have a chance of winning. Candidates with the money to hire professional staff investigate their opponents and provide journalists with negative information if they find it. From the party's perspective, it is better to have this information before the nominee is selected rather than afterward when it would be too late to select a different nominee for the general election.

What matters for both the competitiveness of the race and for democracy is that voters have quality information about the candidates they are covering. The impact of information on a nominating election relates to the balance of information about the candidates. When voters have information about one candidate and little information about other candidates, then voters face a choice between voting for what they know versus going with a candidate they know little about. As a campaign becomes more competitive, voters are more likely to know more about multiple candidates, which empowers voters to select among the candidates.

Media Coverage of Presidential Nomination Campaigns

Political communication has evolved tremendously since the 1970s. The emergence of 24-hour cable news programs, the proliferation of specialized news programs on cable and network television, and the emergence of the internet have affected how the news media cover campaigns. The greater competition for market share puts pressure on print and television news programs to cover topics that generate readership or ratings. Presidential nomination campaigns are hot topics as long as the competition is intense and the nomination remains in doubt. The news media often overstate the degree to which the outcome is in doubt to keep the story interesting to their audiences. News coverage of political campaigns also has become more sensationalistic.

By far, the dominant frame in campaign news coverage is what is called the game schema. A frame is a conceptual perspective that guides the selection of information for coverage and the interpretation of that information. Journalists view politics as a strategic game between politicians competing for advantage.[4] Journalists' concern with the game leads them to pay more attention to the horse race, campaign strategy, and campaign tactics than they do to substantive policy issues or candidate qualifications.[5] Coverage of the *horse race*—who is ahead, behind, gaining, or fading—affects almost all aspects of campaign coverage.

Issues and policies tend to be covered in terms of their impact on the horse race rather than as substantive matters with complex trade-offs. For example, a candidate's position on an issue like immigration is covered in terms of how it will affect the candidates' support rather than in terms of the advantages and disadvantages of that position for the country. The amount and tone of the coverage given to individual candidates reflects journalists' expectations of a candidate's chances of winning the race and closely follows a candidate's support in national polls.[6] Candidates receive more coverage when they have more support in polls and their news coverage changes when their support increases or decreases. The tone of campaign news—how favorable or unfavorable the coverage is to a candidate—also follows the candidates' fortunes in the polls, though scandals, blunders, and other conflicts lead to greater variability in this aspect of news coverage.[7]

Candidates at the top of the polls receive relatively more favorable coverage since reporters focus on favorable candidate attributes and campaign skills to explain their lead. These candidates also are subject to media scrutiny, but at least they get some favorable coverage. Candidates who are trailing in the polls receive disproportionately more negative coverage as journalists and commentators explain why they are trailing in the polls. Often this news coverage focuses on trailing candidates' personal and policy shortcomings or problems with their strategy or campaign organization.[8] Even columnists and cable TV commentators—who are decidedly more ideological in their commentary—are influenced by candidates' position in the polls.[9] When Mitt Romney was trailing in polls during the 2012 campaign, for example, conservative newspaper columnists and FOX news commentators were critical of Romney's inability to rally the base. That kind of criticism declined after the first presidential debate when Romney surged in the public opinion polls. All candidates get coverage that is critical. What differs among candidates is that those at the top of the polls also receive favorable coverage while those at the bottom get very little favorable coverage in the news.

Candidates who gain momentum—those who perform better than expected in the horse race—attract bandwagon coverage. Bandwagon coverage gives a candidate more media coverage that is somewhat more positive since journalists and commentators focus on positive aspects of the candidate to explain their rise in the polls. Bandwagon coverage focuses on and probably contributes to a candidate's momentum in and at the polls. Greater media coverage increases name recognition (especially for lesser-known candidates), which in turn may increase perceptions of a candidate's viability and ultimately increase a candidate's ability to raise funds and attract supporters.[10] Momentum, however, is difficult to sustain since it raises the bar for subsequent primaries. Once they demonstrate success, the candidate is expected to match or beat that success in order to continue to receive favorable coverage. Candidates who decline in or at the polls receive "losing ground" coverage characterized by declining coverage that is more critical as journalists focus on why a candidate is fading.[11] As a result, media coverage expedites the winnowing of candidates fading in or at the polls. It is very hard for these candidates to rebound when their fundraising is drying up and they are receiving criticism in the news media.

What the media say and write about the candidates has some potential to influence primary voters' judgments about the candidates.[12] But the media do not tell people what to think as much as they tell people what to think *about*.[13] The main effect of the news media is an agenda-setting effect. An agenda is a set of items that people are thinking about. The news media—through decisions about which candidates and issues will be covered—influence which candidates will be visible to potential donors and voters. Caucus and primary voters do not evaluate all of the candidates who are seeking the nomination. Instead they focus on the top candidates who are attracting a lot of attention and who are seen as viable candidates. By covering some candidates more than others, the news media play a huge role in telling prospective voters which candidates they should be considering when they go to vote.

Media coverage is especially important early in the nominating campaign, when voters are least informed about the candidates.[14] Media coverage gives candidates visibility, name recognition, and prestige—necessary conditions for being perceived as a viable candidate, which is important for attracting supporters and raising campaign funds.[15] Candidates who fail to attract media coverage are effectively off the radar screen and they are not given serious consideration by many potential caucus and primary voters.

Media images of the candidates appear to influence at least some primary voters' decisions by altering their attitudes about the candidates or their calculations about the candidates' chances of gaining the nomination.[16] The media also prime the audience to think about certain values or policies when they evaluate candidates.[17] Since candidates bring different experiences and mixes of policy expertise, agenda-setting and priming can have strong effects on which candidates will gain an advantage among party activists. Candidates benefit when they receive news coverage and when journalists pay attention to the issues emphasized by a candidate. Candidates are disadvantaged when the media ignore them or when journalists pay attention to the issues and policies emphasized by other candidates.

Because media coverage is so important, candidates adjust almost all aspects of their campaign, from the location and scheduling of events to the content and delivery of speeches in order to maximize the amount and quality of their free media coverage.[18] Candidates use professional staff to help manage the media to get coverage on the candidate's terms and to minimize the damage of unflattering coverage. Some candidates, like Pat Buchanan (1996 Republican) or Al Sharpton (2004 Democratic), have been able to run campaigns on very little money because of their ability to get substantial news media coverage above what would be indicated by polls. It is worth noting, however, that candidates who play to the sensationalist elements of the news media do not seem to gain much of an enduring advantage.

Candidates seek to use the news media to become known and to reinforce the message that they are communicating to prospective caucus and primary voters. A candidate's campaign efforts are reinforced and strengthened when their messages and themes are conveyed in the media. A candidate's campaign efforts are undercut when the news media send a countervailing message. Promoting a politician is quite different from promoting a commercial product like a car or soft-drink. Companies can promote their product with sound and visuals to create a desired emotional effect among a targeted audience. Companies do not have other companies trashing their product. Kia does not have to worry about Toyota running ads criticizing the quality of Kia's cars. Candidates promote themselves in an environment in which skeptical reporters scrutinize their claims and their campaign rivals attack them. The news media communicate these attacks to the broader viewing or reading audience. Most blogs and social media posts involve commentary of subjects covered in the news media.

Information that goes against the candidate's message can undermine its strength and appeal. For example, Hillary Clinton's 2008 nomination campaign strategy involved trying to create an aura of inevitability— a strategy common to front-runners before the caucuses and primaries. The news media, however, often mentioned polls indicating that Clinton had high negatives (unfavorable views) in public opinion polls. Hillary's opponents for the nomination emphasized the limits to her appeal, so that voters would view her as less electable and therefore would take a look at them instead. This kind of countervailing message communicated through the news media can limit the efficacy of a candidate's campaign.

It is worth noting that money and news media coverage are probably less consequential for determining the winner of presidential nominations than is commonly believed, especially by losing candidates who often blame their woes on a lack of money. News media coverage certainly matters. Candidates need visibility and no candidate has enough money to get the exposure that the news media provide. The news media matter most for lesser-known candidates who are long shots to win the nomination. These dark-horse candidates face a dilemma. The perceptions that make them dark-horses also lead the media and campaign donors to ignore them. These candidates usually lack the money needed to run an advertising blitz big enough to become known and they usually cannot raise the money until they are better known. For these candidates, competing for the support of party activists and party identifiers means campaigning through the news media—what political consultants call *earned media* as opposed to paid media or advertising. But the news media tend to follow the polls rather than cause them to change, so candidates must have appeal among party activists to get the news media coverage in the first place. If candidates fail to get news coverage or fail to raise money, it often is because they lack enough appeal with party identifiers to be taken seriously by journalists and donors.

The emergence of niche-market news media has the potential to destabilize patterns of campaign coverage that emerged during the 1960s to the 1990s. Niche-market news media seek to gain an audience by providing content desired by that audience. In the case of political news, niche media seek an audience of liberals or conservatives by telling these viewers, listeners, or readers what they want to hear. Fox News came into being in 1996, with MSNBC following later. These cable news programs offer content that is ideologically biased—supportive of conservative (Fox) or liberal (MSNBC) politics and policy because that is what conservative and liberal citizens want to see or hear.

Psychologists have long known that people pay attention to and give greater credibility to information that is consistent with their beliefs, while ignoring or discounting the credibility of information that conflicts with their existing beliefs. The stronger or more firmly held a person's beliefs are, the more they engage in selectively perceiving information in their environment. The niche news media on cable TV and on the internet take advantage of this fact to carve out an audience—and thus ensure a certain level of profitability because they have an attentive audience as long as they provide that audience with the kinds of information that they want. When it comes to politics, this means providing ideologically biased information. Ideologically defined niche news media like Fox News and MSNBC present a one-sided view of politics because ideological viewers will tune in.

The implications for politics, however, probably reinforce the political polarization that has emerged over the past thirty years. Conservative and liberal party activists increasingly can get information that fits with their own predispositions and they can ignore information that conflicts with their predispositions.[19] Both Democratic and Republican Party activists get a distorted view of the world in which their side is portrayed as good and the other side as bad. Such people are less likely to recognize that their own policy preferences are those of a special interest and that both parties seek to advance agendas that benefit special interests. Instead, party activists attend to information that equates their specialized interests with that of the national interest or public good.

This is not new. Columnists and editorials in the op-ed pages of newspapers have long written this kind of commentary. Conservative columnists in newspapers are critical of Republican candidates or office holders when they do not follow conservative principles. Liberal columnists criticize Democratic candidates for doing the same with respect to liberal policies. What differs is that this kind of editorial coverage is now the "news" coverage of political campaigns. Journalists and news reporters offer commentary on the campaign rather than reporting about the campaign. The result is that only a small portion of political news coverage actually is of the candidate offering ideas and arguments. Mostly candidates are shown giving a very brief seven or eight second sound-bite that is then discussed or commented on by the reporter. This kind of editorial journalism blends fact and opinion in ways that is often undifferentiated by readers or viewers. Opinions are readily taken as fact by audiences who accept such content when it aligns with their own political predispositions.

Within presidential nomination campaigns, the implication is that more liberal candidates (in the Democratic Party) and more conservative candidates (in the Republican Party) get favorable coverage on these niche media more than has been the case for much of the past century. Further, these niche news media push candidates to adopt relatively extreme positions. Republican presidential candidates are criticized for having moderate positions while Democratic presidential candidates are criticized by such media for not espousing liberal positions. Candidates seeking to be seen and heard by the partisan audience have an incentive to adopt positions and speak to the more extreme preferences of the partisan audience.

The Internet and Social Media: Micro-Campaigning

The emergence of the internet and social media have also changed the strategies and tactics of presidential candidates, affecting both activists and others trying to influence the nomination as well as enabling presidential campaign organizations to target individuals for messages and mobilization. The national political parties are coalitions of diverse groups and individuals with interests and preferences across a range of issues and policies. Sophisticated campaigns are engaged in identifying the voting propensities and political preferences of numerous subsets of the electorate so that a campaign can give these diverse groups a common direction. Identifying prospective supporters has become increasingly sophisticated, with substantial advances in data mining and analytics—a multifaceted approach to market analysis that combines inductive and deductive methods of social research. Political organizations gather, score, and model demographic and behavioral data as well as attitudinal and opinion data to find voters most likely to support a candidate as well as undecided swing voters. Campaign strategists, fundraisers, and messaging experts use the information to better persuade voters by identifying and communicating the right messages through targeted and personalized direct mailings, emails, phone calls, or door-to-door canvassing.

The potential of the internet as a campaign organizing mechanism was advanced most notably by the Howard Dean campaign for the 2004 Democratic nomination. The Dean campaign sought to create something of a sociopolitical phenomenon by encouraging supporters to get together for rallies. The Dean campaign recruited and organized supporters through Meetup.com, an early social media forum.[20] The buzz created by the tactic drew traditional media coverage and created energy among Dean supporters, who were intensely opposed to the War in Iraq. The Dean campaign, however, would not be sufficient to gain the nomination. The big advance in internet and social media occurred with Barack Obama's nomination campaign in 2008.

The first Obama presidential campaign had an impressive market research operation that enabled them to engage in detailed messaging through TV and social media. The campaign built a donor base of 1.5 million people during the nomination phase of the campaign.[21]

In addition to the massive fundraising operation, which outpaced any prior presidential nomination campaign, the Obama organization used the internet interactively and collaboratively to identify, train, and monitor the progress of grassroots organizers at the precinct level in competitive states. The campaign used the internet and social media in particular to gain a huge organizational advantage in caucus states, which Hillary Clinton's campaign largely ignored.[22] His online supporters created more than 30,000 events to promote his candidacy before the end of the caucus and primary season.[23] The campaign also reached out interactively to supporters for feedback on and contributions to the campaign's policy pages. It used email extensively as a cheap, almost unlimited volume, messaging system to update and feed party activists' appetite for information about their candidate and to counter attacks by opponents. The campaign continuously streamed new material through videos posted to YouTube, messages sent through Twitter, postings on Facebook, and messaging on email. The result was a multi-media, high-volume campaign targeting party activists in ways that had never before been used on the scope or with the success of the Obama campaign.

After the campaign, the Obama team kept the operation going with extensive research on voters. The key additions for the 2012 campaign included more extensive use of market analytics, which involves using massive amounts of data on people to discern patterns of behavior that are in turn used to identify potential supporters, and personalized messaging to vastly more and more narrowly defined subsets of the population.[24] The campaign's "analytics" department identified key constituencies, their concerns, and their patterns of media usage—not just at the group level, but to the extent that they were able to identify individuals for targeting. This information helped the Obama campaign to craft messaging and to reach targeted populations more efficiently through a multi-media communications approach and through extensive personal contact. The campaign spent months developing and testing messages with small samples and then went live with bigger blasts of information and messaging that had proven to resonate with targeted audiences. Further, the campaign went beyond prior efforts by integrating the market research and messaging with the grassroots operation to get out the vote. Activists on the ground were given apps that would include the messaging to be used for interaction with specific individuals in targeted neighborhoods.

The emergence of technologies and the fusion of big data—merging consumer data with political data—can now be used to supplement the mass media appeals to party activists. This kind of micro-targeting is made possible by technology and the emergence of vast amounts of information on individuals—from consumer behavior to people's online activity—which is fused with political data from various sources. This information is used to identify individuals who might be persuaded to support a campaign and for the development of messages that appeal to that person. The campaigns use data on social networks to identify individuals who already support the campaign and contact that supporter to reach out to the targeted individual who has not yet signed on as a supporter. The revolutions of technology promise to make campaigns much more intensely focused on campaign activists than has been done in the past.

For their part, party activists and groups aligned with the political parties have used the internet to engage in extensive communication and information sharing.[25] These information networks increasingly allow activists and groups to weigh in on candidates and the campaigns in an evaluative sense, as occurs on blogs, Facebook postings, Twitter, email blasts, YouTube videos, and more. The point is that activists and groups have greater potential than ever before to communicate and coordinate among themselves as to which candidate should be selected as the presidential nominee of the political party. Perhaps more than any other time in history, there is capacity for group leaders and party activists to communicate their perceptions of candidates to other activists around the country. The medium thus has expanded the voice and influence of groups aligned with the political parties that seek to influence nominations. Candidates have more incentive than ever before to cater to activist groups that seek to influence the political parties from within.

The internet thus is a tool that democratizes the nomination process in the sense of extending the participatory franchise to an ever-increasing number of potential participants. More people than ever can put in the public domain their opinions and views on candidates and policy. In an increasingly digitally networked world, the reach of these opinions depends on the scope of an individual's connections in the networks that form the modern political parties.

A big question going forward will be whether the extensive communication possibilities will be used to tear apart candidates or to rally support toward the banner of one of the candidates in the race. Political parties are diverse and so many different demands made visible may make it harder for activists to find a candidate that is acceptable to all. Virtually everything a candidate says and does in public life can be recorded, pulled out of context, and recast for effect in a political attack ad. Candidates no longer have to concern themselves with a few dozen reporters scrutinizing their words. And opponents are not the only ones attacking candidates. Technology enables thousands of party activists to create and disseminate commentary in script and video format on YouTube.

It seems likely that the extensive communication of the blogosphere and Twitter-sphere detracts from the aura of candidates more than it helps one of them draw support from across the political party. Almost anyone with strong convictions about public policy can now weigh in on a presidential nomination campaign. While the vast majority of commentary by political activists will be seen or heard by a handful of their friends, it is possible that some commentary or video upload goes viral—rapidly diffusing through the internet and being picked up by the broadcast media.

The potential atomization of political communication among vast numbers of party activists and groups makes it harder for candidates and their staffs to get party activists moving in the same direction to support a particular candidate. There is an enormous amount of noise in recent nomination campaigns. The power of well-funded groups, operating within the structure of a political party but not beholden to party objectives, would seem to have enormous potential in the era of the internet and specialized media. Whether the internet enables greater coordination or fragmentation of constituencies within a political party remains to be seen. The proliferation of voices on social media, however, would seem to make it harder for presidential candidates to get large numbers of party activists moving in the same direction.

Notes

1 Patterson, 1980, *The Mass Media Campaign*; Michael J. Robinson and Margaret Sheehan, 1983, *Over the Wire and on TV: CBS and UPI in Campaign '80*, New York: Russell Sage; Wayne P. Steger, 2002, "A Quarter Century of Network News Coverage of Candidates in Presidential Nomination Campaigns," *Journal of Political Marketing*, 1(1): 91–116.
2 Graham P. Ramsden, 1997, "Media Coverage of Issues and Candidates: What Balance Is Appropriate in a Democracy," *Political Science Quarterly*, 1(1): 65–96.
3 Lipsitz, 2011, *Competitive Elections and the American Voter*.
4 Thomas E. Patterson, 1993, *Out of Order*, New York: Alfred A. Knopf, 57.
5 Christopher F. Arterton, 1984, *Media Politics: New Strategies of Presidential Campaigns*, Lexington, MA: D.C. Heath.
6 Patterson, 1993, *Out of Order*.

7 Henry E. Brady and Richard Johnston, 1987, "What's the Primary Message: Horse Race or Issue Journalism," in *Media and Momentum*, Gary R. Orren and Nelson W. Polsby (eds.), Chatham, NJ: Chatham House, 127–186.

8 Robinson and Sheehan, 1983, *Over the Wire and on TV*, 123.

9 Wayne P. Steger, 1999, "Comparing News and Commentary Coverage of the 1996 Presidential Nominating Campaign," *Presidential Studies Quarterly*, 29(1): 40–64.

10 Richard Joslyn, 1984, *Mass Media & Elections*, Reading, MA: Addison-Wesley, 129; Larry Bartels, 1985, "Expectations and Preferences in Presidential Nominating Campaigns"; Brady and Johnston, 1987, "What's the Primary Message"; Patterson, 1993, *Out of Order*, 117–118; Diana C. Mutz, 1997, "Mechanisms of Momentum: Does Thinking Make It So?" *Journal of Politics*, 59(1): 104–125; Patterson, 1980, *The Mass Media Election*, 43–48; Emmett H. Buell, 1987, "Locals and Cosmopolitans: National, Regional, and State Newspaper Coverage of the New Hampshire Primary," in *Media and Momentum*, Gary R. Orren and Nelson W. Polsby (eds.), Chatham, NJ: Chatham House.

11 Patterson, 1993, *Out of Order*, 119–120.

12 Popkin, 1991, *The Reasoning Voter*.

13 Shanto Iyengar, 1991, *Is Anyone Responsible: How Television Frames Political Issues*, Chicago: University of Chicago Press; Maxwell E. McCombs and Donald L. Shaw, 1972, "The Agenda Setting Function of the Mass Media," *Public Opinion Quarterly*, 36(2): 176–187; Diana C. Mutz, 1989, "The Influence of Perceptions of Media Influence: Third Person Effects and the Public Expressions of Opinions," *International Journal of Public Opinion Research*, 1(1): 3–33.

14 Cliff Zukin and Scott Keeter, 1983, *Citizen Learning in Presidential Nominations*, New York: Praeger.

15 Robert L. Peabody, Norman J. Ornstein, and David W. Rhode, 1976, "The United States Senate as a Presidential Incubator: Many Are Called but Few Are Chosen," *Political Science Quarterly*, 91(2): 237–258.

16 Bartels, 1988, *Presidential Primaries*; Brady and Johnston, 1987, "What's the Primary Message"; Paul R. Abramson, John H. Aldrich, Phil Paolino, and David W. Rhode, 1992, "Sophisticated Voting in the 1988 Presidential Primaries," *American Political Science Review*, 86(1): 55–69.

17 Shanto Iyengar and Donald R. Kinder, 1987, *News that Matters: Television and American Public Opinion*, Chicago: University of Chicago Press; Jon A. Krosnick and Donald R. Kinder, 1990, "Altering the Foundations of Support for the President through Priming," *American Political Science Review*, 84(2): 497–572.

18 Herbert B. Asher, 1984, *Presidential Elections and American Politics*, 3rd ed., Homewood, IL: Dorsey Press; Audrey A. Haynes, Paul-Henri Gurian, Michael H. Crespin, and Christopher Zorn, 2004, "The Calculus of Concession: Media Coverage and the Dynamics of Winnowing in Presidential Nominations," *American Politics Research*, 32(3): 310–337.

19 Natalie Jomini Stroud, 2008, "Media Use and Political Predispositions: Revisiting the Concept of Selective Exposure," *Political Behavior*, 30(2): 341–366; Natalie Jomini Stroud, 2010, "Polarization and Partisan Selective Exposure," *Journal of Communication*, 60: 556–576; Natalie Jomini Stroud, 2011, *Niche News: The Politics of News Choice: The Politics of News Choice*, Oxford: Oxford University Press.

20 Gary Wolf, 2003, "How the Internet Invented Howard Dean," www.wired.com/wired/archive/12.01/dean.html.

21 Sarah Stirland, 2008, "Obama, Propelled by the Net, Wins Democratic Nomination," www.wired.com/threatlevel/2008/06/obama-propelled/.

22 Christine Williams and Girish Gulati, 2008, "What Is a Social Network Worth? Facebook and Vote Share in the 2008 Presidential Primaries," Paper presented at the American Political Science Association annual meeting, Boston, MA.

23 Stirland, 2008, "Obama, Propelled by the Net, Wins Democratic Nomination."

24 Steger, 2013, "A Transformational Presidential Campaign."

25 Gregory Koger, Seth Masket, and Hans Noel, 2009, "Partisan Webs: Information Exchange and Party Networks," *British Journal of Political Science*, 39(3): 633–653.

6

COMPETITION IN POST-REFORM PRESIDENTIAL NOMINATIONS

How leaders are selected and who does the selecting reflects power in a political system. Democratic political systems are premised on the notion that citizens are empowered to select their political leaders. However, voters must have choices among candidates to exercise that power. At face value, the reforms of the 1970s made presidential nominations more democratic by enabling many more citizens to participate in the selection of the party nominees. The reforms also enabled more candidates to seek the nomination so there are more candidates to choose from.

But, as we have seen, the early phase of the nomination campaign provides opportunities for party stakeholders to coordinate their support for a candidate and forge a winning coalition even before citizens cast ballots in the caucuses and primaries.[1] When a particular candidate has a huge lead because party stakeholders have unified behind that candidate, then there is less effective competition among candidates once the voting begins.[2] Voters basically have a vote of confidence (or no confidence) in the candidate preferred by party stakeholders. Even then, voters will have a hard time finding a viable alternative to the front-runner. However, if party stakeholders are undecided or divide their support among several candidates, then voters become the arbiters of the nomination competition.

Competition among political organizations and leaders provides citizens with the opportunity to make meaningful choices in elections.[3] Evaluating the competitiveness of the race and the number of viable candidates that voters choose among informs us about who exercises power in a political system and thus about how democratic the leadership selection process is. If nominations are effectively decided during the invisible primary, then there will be little competition in the primaries and voters will essentially have one viable candidate that they can accept or reject. Caucus and primary voters usually go along with the choice of the party stakeholders when stakeholders reach agreement on who should be nominated.

When party stakeholders remain divided or uncommitted during the invisible primary, the race will be competitive going into the caucuses and primaries where voters will have choices among two or more viable candidates. In this scenario, there is more competitive balance among the candidates who are more evenly matched in their support from party stakeholders, in their fundraising, and in the coverage they receive from the news media. There is more uncertainty about who will win the nomination, and campaign momentum has more potential to affect the outcome.

The question is how to observe and measure the competitive balance of the race and the number of viable candidates that voters have to choose among. The number of viable candidates—those with a reasonable chance of winning the nomination—and the competitive balance of a nomination race have been measured with the Hirschman-Herfindahl Index (HHI).[4] This statistic has been used to measure all kinds of competition, from the competitiveness of professional sports leagues to the competitiveness of markets for specific industries. It works well to assess the competitiveness of elections in which there are multiple candidates, and it can be used to measure the number of effective or equivalent options available to voters—that is, it provides a sense of the number of candidates with a similar chance of winning. The reciprocal of the index (1 divided by the HHI) provides a measure of the number of equivalent or in this case viable candidates in a multi-candidate election.[5]

The measure is better than just counting the number of candidates on the ballot because many of the candidates do not have a realistic chance of winning. Some serve mainly to give voters a chance to express a symbolic statement. Consider, for example, the 2012 Republican nomination race which included U.S. Representatives Michelle Bachman (MN) and Ron Paul (TX), former Ambassador John Huntsman (UT), Governors Tim Pawlenty (MN) and Rick Perry (TX), former Senator Rick Santorum (PA), Former Speaker of the House of Representatives Newt Gingrich (GA), Former Governors Mitt Romney (MA) and Buddy Roemer (LA), and businessman Herman Cain (VA). Counting the number of candidates would make this race seem like there were a lot of choices available to voters—most were on the ballots in multiple states even after they had withdrawn from the race. But few of these candidates had a chance of winning and most probably did not receive serious consideration by voters. Pawlenty and Cain quit the race during the invisible primary. Bachman, Roemer, Huntsman, and Perry dropped out of the race after receiving very few votes in the Iowa caucus and the New Hampshire primary. Ron Paul had support from only a small minority of Republicans. Gingrich and Santorum each had serious liabilities and few observers thought they had a realistic chance of winning the nomination much less the general election. Even the eventual nominee—Romney—wasn't particularly popular among conservative Republicans. His main advantage may have been that he was widely seen as the candidate with the best chance of defeating Barack Obama in the general election. In a campaign like this, simply counting the number of candidates on the ballot is misleading. The inverse of the HHI for the 2012 Republican primaries is 2.76—or essentially more than two and almost three viable options for voters to choose among. The resulting figure is probably pretty close to the number of candidates considered by voters in many if not most of the primaries.

The HHI also can be normalized to create a measure of the competitiveness of the race, controlling for different numbers of candidates in different election years.[6] The normalized HHI is interpreted under the assumption that low concentration scores indicate more competition (no candidate dominates). High scores indicate less competition— a score of 1 indicates no competition. The Anti-Trust division of the U.S. Department of Justice (DOJ) uses the normalized HHI to determine whether a merger of corporations would excessively restrict market competition. The DOJ generally considers markets in which the HHI is between .15 and .25 to be moderately competitive and scores above .25 to indicate little competition.[7]

To illustrate how the index works, consider the 2012 Democratic and Republican nomination races. On the Democratic side, President Obama sought renomination. Although the national media showed that Obama ran unopposed, there were actually five other candidates who gained votes in some of the Democratic primaries. While six candidates obtained votes in Democratic primaries, the index shows that there were only 1.02 viable candidates in the race. The statistic also shows that this was not a competitive race with a normalized HHI score of .94—close to the level of 1.0 that indicates no competition at all. Although Obama was not the only candidate on the ballot, he was the only one with any chance of winning the Democratic nomination. On the Republican side, there were 10 candidates who received at least a tiny share of the vote in some primaries, 4 of which who received more than 1% of the vote in all of the primaries taken together. The index indicates that there were 2.75 viable candidates in this race, indicating that Republican primary voters did have choices among candidates, though less than the number that appeared on the ballot. The normalized HHI is .15 for these primaries, indicating that this was a moderately competitive race.

The important thing is that the statistic allows us to compare the competitiveness of a race and the number of viable choices available to voters across campaigns with different numbers of candidates appearing on the ballot. A low level of competition in the primaries and a low number of viable candidates indicates that there was substantial degree of agreement about which candidate should be nominated. This occurs when party stakeholders and party identifiers have already figured out who they will nominate even before the caucuses and primaries begin. If nominations are not resolved by the end of the invisible primary, there will be more competition in the primaries. In this scenario, we should see two or more viable candidates for voters to choose among and a normalized HHI statistic below the .25 threshold, indicating a race with reasonably balanced competition among candidates.

To preview, half of the nominations between 1972 and 2012 have had low levels of competition, indicating a selection process that produces substantial agreement about the nominee even before the voting begins. The other half of the races have exhibited moderate to high levels of competition, which indicates a failure to resolve the nomination during the invisible primary. While the overall picture looks like a nomination process that sometimes determines the outcome before voters cast ballots, there are differences across time, between the two political parties, and between races involving an incumbent president seeking renomination and "open" nominations—races without an incumbent.

Figure 6.1 shows the number of viable candidates in Democratic and Republican presidential primaries in the post-reform era (1972 to 2012). Figure 6.2 shows the competitive balance among candidates for each nomination race. These figures show three general patterns. First, nominations involving a president seeking a second term are usually not competitive compared to open nominations. Races with an incumbent nomination have an average normalized HHI of .63, indicating very little competition on average. Open nominations have an average normalized HHI of .23, indicating moderately competitive races. Second, nominations of the 1970s were typically more competitive than those occurring since then (see Figure 6.1). Third, Republican nominations have generally been less competitive than have Democratic

87

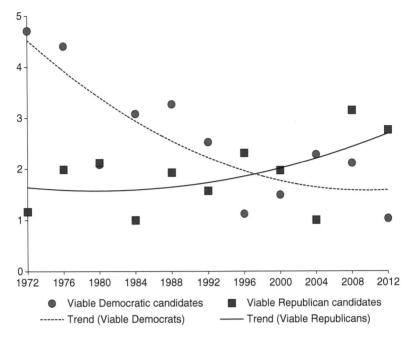

Figure 6.1 Number of Viable Candidates in Presidential Primaries, 1972 to 2012

Source: Author's calculations using a Hirschman-Herfindahl Index of candidate shares of the cumulative vote across all presidential primaries in each year.

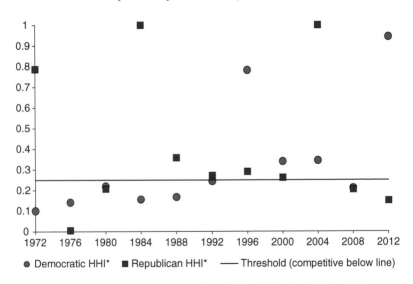

Figure 6.2 Competition in Democratic and Republican Primaries, 1972 to 2012

Source: Author's calculations using HHI*, a *normalized* Hirschman-Herfindahl Index of candidate share of the vote across all presidential primaries.

nominations. The average normalized HHI for open Republican races has been .24, which means these races have been moderately competitive on average, but just barely so. The score for open Democratic nominations has been .21, indicating that these races have been somewhat more competitive on average. We will address each point in turn, beginning with the distinction between incumbent races and open races.

Incumbent Renominations versus Open Nominations

Presidents rarely face a serious challenger from their own party when they seek renomination to run for a second term. Campaigns with an incumbent president seeking renomination usually have fewer viable candidates than "open" nominations—those nomination without an incumbent seeking reelection. Open nominations in the post-reform era have had an average of 2.7 viable candidates in the primaries, compared to an average of about 1.4 viable candidates in the primaries when an incumbent president sought renomination (see Figure 6.1). Presidents Nixon (1972), Reagan (1984), Clinton (1996), Bush (2004), and Obama (2012) faced no serious opposition to their renomination. Presidents Ford (1976) and Carter (1980) each faced a serious challenger. President George H. W. Bush also faced a minor challenge to his renomination in 1992.

The race between Gerald Ford and Ronald Reagan in 1976 involved the most balanced competition of any presidential nomination in the post-reform era. The circumstances of this race were highly atypical of incumbent renomination races. Ford became president after Richard Nixon resigned in 1974 to avoid being the first president to be removed from office. Nixon Administration officials and his campaign's operatives had engaged in a number of illegal activities in what became known as the Watergate Scandal. Vice presidents like Ford who become the president by means of succession are usually challenged if they seek the nomination of their party. Of the nine vice presidents *ever* to succeed to the presidency on the death or resignation of the elected president, only Theodore Roosevelt in 1904 was unchallenged for his renomination.[8] Ford was even more vulnerable than most since he had not even been elected vice president. He was appointed to the vice presidency after Spiro Agnew resigned in the face of a bribery and tax evasion scandal. In this context, it is not surprising that Ford faced a tough, competitive challenge to his nomination in 1976.

Beyond that, however, the 1976 Republican presidential nomination appears to have been a battle for the heart and soul of the Republican

Party. Ford sought the support of the moderate "establishment" wing of the Republican Party, which won in 1976 (more on this in chapter eight). Reagan was the preferred candidate of conservative Republicans. Reagan lost in 1976, but won the Republican nomination in 1980— having a profound impact on the groups forming the Republican Party coalition.[9] Between 1976 and 1980, Reagan's supporters worked hard to build support for his candidacy, developing a leadership political action committee (PAC) to raise funds for (and curry favor among) Republican candidates. Reagan's supporters also sought to appeal to evangelical Christians who formed a group called the Moral Majority headed by Jerry Falwell. Reagan's 1980 nomination campaign essentially expanded the Republican Party coalition to bring in what became known as the Christian Right (more on this in chapter eight). While there was some competition in the 1980 Republican nomination, opponents of Reagan could not agree on which of the moderate alternatives should be nominated. From Reagan's nomination in 1980 until George W. Bush's renomination in 2004, Republican nominations were less competitive (see Figure 6.2). This suggests that the Republican Party coalition exhibited substantial stability between 1980 and 2004. The Republican Party coalition appears to have started to fragment by 2008, as will be discussed more in chapter eight.

Jimmy Carter was a Southern Democrat who was more conservative than most members of his own political party. Carter won the 1976 Democratic nomination in part because the numerous liberal Democratic candidates divided the votes of liberals.[10] Carter's position as a moderate Democrat made him vulnerable to a challenge by a candidate from the dominant, liberal wing of the Democratic Party.[11] The Democratic Party coalition was highly divided in the 1970s, which had the effect of making nominations more open and competitive, as will be discussed more in chapter eight. Carter also benefitted from the uncertainty that resulted from the rules of the nominating campaign which were (at that time) still relatively new. Carter adopted a strategy that was innovative at the time, concentrating his campaign efforts in the Iowa caucus in order to gain momentum by beating expectations.[12]

In 1980, Carter was challenged by Senator Ted Kennedy—the younger brother of President John F. Kennedy and former Attorney General Robert Kennedy. Kennedy had been the front-runner in pre-election year polls before the 1972 and 1976 Democratic nominations. Although he was the party favorite in both years, he did not run after a scandal in which he fled the scene of a car accident at Chappaquiddick Island in which his passenger died (a young woman named Mary Jo Kopechne). By the late 1970s, however, Kennedy had reassumed a leadership role in the U.S. Senate—a position he used to challenge Carter leading up to the 1980 nomination. While Carter's renomination race was competitive, it is important to realize that all of the Democratic nominations were at least moderately competitive until 1996. The "New Deal" coalition that made the Democratic Party the majority party in the country from the 1930s to the 1960s was fragmenting rapidly in the 1970s. The Democratic Party coalition was highly divided from the 1960s through the 1990s.[13] As we will see in chapter eight, a divided party coalition has the effect of making nominations more competitive.

Notably, both of these challenges failed to stop the renomination of the incumbent president. Incumbent presidents are exceedingly difficult to beat when they seek renomination to be their political party's presidential candidate. The fact that these presidents were challenged so vigorously indicates that the reformed nomination process can serve as a referendum on the president for a party's activist base to signal their dissatisfaction with the president's leadership. That the two competitive renomination campaigns occurred during the 1970s is notable for two reasons. First, the reformed nomination process was relatively more open in the 1970s, which invited more competition for presidential nominations of both parties. A challenger could hope to appeal to party activists and party identifiers dissatisfied with the president. After the 1980 election cycle, insiders at the national and state levels implemented a number of reforms that had the effect of limiting the extent of competition. Second—and probably more important—both political parties were experiencing a lot of coalition change during the 1970s. The realignment of the party coalitions from the late 1960s to the early 1980s made nominations less predictable and more competitive. Both parties gained more coalitional stability by the late 1980s, at least in comparison to the situation in the 1970s. The unity and stability of the party coalitions has a big impact on the competitiveness of presidential nomination campaigns.

Compared to incumbent renomination races, open nominations without an incumbent president typically are more competitive and voters choose among a larger number of viable candidates. Primary voters have had an average of 2.7 viable candidates to select among in open nomination races between 1972 and 2012 (see Figure 6.1). This compares to an average of 1.4 viable candidates in incumbent renomination races during this time period. Open nominations have an average normalized HHI score of .23—compared to a score of .63 for incumbent renominations—indicating a substantially more competitive situation. Open nominations usually attract multiple candidates who are able to obtain the campaign funds and media coverage to compete for the support of voters in the primaries, as will be discussed in the next chapter. Of the 14 open nominations from 1972 to 2012, only 5 races have been relatively uncompetitive during the primaries. Three of these were Republican nominations in 1988, 1996, and 2000. The 1988 Republican nomination featured Vice President George H. W. Bush who had substantial support from party insiders as well as a substantial fundraising advantage (as will be discussed more in the next chapter). The 1996 race featured Republican Senate leader Bob Dole who had been Ford's vice presidential candidate in 1976 and the runner up in the 1988 Republican nomination. The 2000 Republican nomination had George W. Bush, who was by far the candidate with the most support from party insiders and donors. The other two relatively uncompetitive open nominations occurred on the Democratic side. The 2000 Democratic nomination featured Vice President Al Gore who had substantial advantages over his only rival, former Senator Bill Bradley. The other less competitive Democratic race occurred in 2004, which may surprise some readers. This race was competitive at the end of the invisible primary with no clear front-runner. However, the strongest rival to Kerry was Governor Howard Dean—a candidate whose campaign imploded with an unexpected loss in the Iowa caucus. Within weeks of this election, most of the candidates had withdrawn from the race, leaving Kerry to run almost unopposed through most of the remaining primaries. In this case, it appears that very strong campaign momentum rather than pre-primary collusion accounted for a relatively less competitive race.

The other nine open nominations featured more competition for the nomination and two or more viable options for voters. That is, it does not seem to be the usual case that party stakeholders are able to coalesce behind a candidate sufficiently to make the caucuses and primaries a foregone collusion. The open Democratic nominations of 1972, 1976, 1984, 1988, 1992, and 2008 all had relatively balanced competition across the primaries taken as a whole. The open Republican nominations of 1980, 2008, and 2012 were similarly competitive through the caucuses and primaries. On average, it would appear that party stakeholders can anoint a front-runner as the "inevitable" nominee during the invisible primary phase of the campaign.

Declining Competition in Presidential Nominations

The competitiveness of presidential nominations has changed over time. The nomination campaigns of the 1970s were highly competitive races with a historically high number of viable candidates for voters to choose among. The competitiveness of nomination campaigns declined after 1980 as the result of forces that affected both political parties.

Perhaps the main reason that the nominations were competitive in the 1970s into the early 1980s was that both parties were in a period of transition as the political parties realigned at the presidential level. The New Deal coalition of the Democratic Party was fragmenting as Southern whites moved away from the party. Over time, white voters in Southern states have come to identify with and support the Republican Party. At the same time, citizens in Northern states, especially those in the Northeastern states, have moved away from the Republican Party and toward the Democratic Party. These geographic and socioeconomic changes in the memberships of the political parties have made both parties more ideologically homogenous. The increasing homogeneity of the party constituencies makes it easier for stakeholders to unify behind a presidential candidate even before the caucuses and primaries begin.[14]

Presidential nomination races also became less competitive in part because the process became less open as party insiders pushed back against the reforms. States began holding their caucuses and primaries sooner and often on the same dates, giving "dark-horse" candidates less time to take advantage of momentum if they win an early nominating election. Candidates in the 1980s and 1990s lacked the time needed to raise money and to build a nation-wide campaign organization. The rising costs of mass media campaigns also had an impact. Candidates who cannot raise money often drop out of the race because they cannot afford to compete in enough primaries to have a chance of winning the nomination. The recent emergence of fundraising online has allowed candidates to overcome the impact of a compressed primary schedule. Thus, the period from 1984 through 2000 limited the potential for dark-horse candidates to break through by gaining momentum in the early contests.

A third factor contributing to less competitive nomination campaigns since the 1970s has been the change in the behavior of candidates and other participants in the nomination process. Candidates learned how to campaign in the system of rules created by the McGovern-Fraser reforms. Since 1976, all of the front-runners have tried to prepare for a national campaign of endurance in the caucuses and primaries with the goal of surviving set-backs and other bumps on the road.[15] Nearly all "dark-horse" candidates have adopted the Jimmy Carter strategy of entering the race early, concentrating resources in one of the early caucuses or primaries (usually Iowa and/or New Hampshire), and hoping to break through.[16] This homogenization of strategies has made the races less competitive overall since candidates more quickly realize when they cannot win. More candidates drop out of the race during the invisible primary than they used to and more candidates drop out after the earliest events. During the 1970s, candidates entered the competition late and stayed in the race longer—often fighting to the very last primary. Far fewer candidates decide to remain in the race if they don't do well in the first few contests. As a result, there are fewer viable candidates across the primaries and less competition throughout the primary season.

Presidential primaries remain competitive as long as there are several viable candidates remaining in the race. The 2004 and 2008 Democratic nominations illustrate the situation that has emerged over time. In 2004, there were eight candidates who entered the Iowa caucus and Howard Dean was a weak front-runner in insider endorsements and national opinion polls. Dean began to fade in the weeks leading up to the Iowa caucus—a contest in which he finished a distant third behind John Kerry and John Edwards. Most of the Democratic candidates dropped out of the race within two weeks of the New Hampshire primary. The result was a race with little competition after the first two or three weeks of the primary season. Similarly in 2008, there were eight candidates for the Democratic nomination. Only two of these candidates remained viable candidates after the initial four contests. The remaining race was a two person competition in which Barack Obama slowly gained more delegates than Hillary Clinton. Although Clinton continued to contest primaries to the end, Obama continued to build momentum until he gained enough delegates to become the nominee.

It is worth noting that the Democratic primaries have had more competitive balance among the candidates from the 1970s until the 1990s. Democratic nominations since Bill Clinton's initial nomination in 1992 have been somewhat less competitive than they were before that time. In contrast, Republican nomination campaigns were relatively uncompetitive from 1980 until 2008. The two parties' nomination campaigns have switched positions with respect to the competitiveness of these races. Democratic nomination campaigns have become less competitive while Republican nomination campaigns have become somewhat more competitive (see Figure 6.1). The reasons for the change will be discussed in subsequent chapters.

Democratic versus Republican Nomination Campaigns

The patterns of candidates and competition in Figures 6.1 and 6.2 show that there have been differences between the two political parties. The difference in the competitiveness of presidential nomination campaigns for the two parties reflects changes occurring across time within each party. These differences are due mainly to: (1) differences in the relative unity of the political parties, and (2) decisions of candidates about entering or dropping out of the race. The Democratic Party was relatively more divided internally in the 1960s and 1970s than it has been in the last decade. In contrast, the Republican Party arguably has become more divided in the past decade than it has been since 1976.

These patterns can be seen in the two trend lines in Figure 6.1, indicating the changes in Democratic and Republican presidential nominations. Democratic presidential nomination campaigns (the dotted line) had a large number of effective candidates and were more competitive in the 1970s and 1980s and have become less competitive with fewer viable candidates contesting primaries since the Clinton years of the 1990s. Republican nomination campaigns (the solid line) were competitive in 1976 but had low levels of competition from 1980 through 2004. The last two Republican nominations have been more competitive as a larger number of viable Republican candidates have remained in the primaries longer than they did in the 1980s and 1990s. Republican candidates seeking the nomination were more likely to drop out of the race if they failed to win an early caucus or primary during those earlier nominations. The result has been more competitive races deeper into the primary season for Republicans while Democratic races have become less competitive on average.

Conclusions

Presidential nominees have been predicted with a great deal of accuracy in some nomination campaigns but not others. Forecasts are accurate when a presidential nomination is effectively settled during the invisible primary.[17] This is what should happen if party stakeholders coalesce sufficiently behind a candidate during the invisible primary. Those races that have not been resolved during the invisible primary have been more influenced by campaign momentum, with the Iowa caucus and New Hampshire primary having the most influence on the outcome. Notably, *all* of the presidential nominations, however, are forecast accurately once the models take into account the effects of campaign momentum gained in the earliest caucuses and primaries. The difficulty forecasting Democratic primary votes reflects the greater competitiveness of Democratic nomination campaigns. Democratic primaries were at least moderately competitive in all of the nominations from 1972 to 1992, including the 1980 race involving President Carter's renomination. Democratic primaries have been less competitive since 1992. This suggests that Democratic Party stakeholders were less able to unify behind a candidate during the invisible primaries before Bill Clinton became president. But even when Democratic stakeholders have coalesced behind a candidate, that candidate has not always won the nomination. The Republican primaries, by comparison, have been competitive only in 1976, 1980, 2008, and 2012. Republican stakeholders were unified in all of their nominations between Reagan's nomination in 1980 and the last two presidential nominations.

The next chapter will examine whether and to what extent party elites, campaign contributors, and party identifiers have coalesced behind a front-runner by the end of the invisible primary. If the nominations that feature party unity by the end of the invisible primary are the same as those that lack competition during the primaries, then there is pretty strong evidence that the critical period of decision has been during the invisible primary. As we will see, the invisible primary has been more critical for selecting Republican presidential nominees than for Democratic presidential nominees. Democratic elites and party identifiers often fail to unify behind a front-runner during the invisible primary— and on occasion have gone in different directions in their candidate preferences. Republican elites and identifiers generally have been more in sync and have unified behind a candidate even before the votes are cast in caucuses and primaries. There is, however, some evidence that the unity of the Republican Party—forged during the Reagan years—is beginning to fragment.

Notes

1 Cohen et al., 2008, *The Party Decides.*
2 Steger, Hickman, and Yohn, 2002, "Candidate Competition and Attrition in Presidential Primaries."
3 Joseph Schumpeter, 1942, *Capitalism, Socialism, and Democracy*, London: Allen and Unwin; David Held, 1987, *Models of Democracy*, Stanford, CA: Stanford University Press.

4 Steger, Hickman, and Yohn, 2002, "Candidate Competition and Attrition in Presidential Primaries."
5 Rein Taagepera and Matthew Soberg Shugart, 1989, *Seats and Votes: The Effects and Determinants of Electoral Systems*, New Haven, CT: Yale University Press.
6 The formula is given by $H = \sum c_i^2$, where c_i = a candidate's share of the vote across all of the primaries and $HHI^* = (H - 1/N) / (1-1/N)$, where N is the total number of candidates in the race. All individual candidates receiving more than 1% of the aggregate primary vote were included in the analysis. The inclusion of candidates with a fraction of a percentage point of the vote has a trivial effect on the HHI scores. HHI^* indicates a normalized Hirschman-Herfindahl Index.
7 www.justice.gov/atr/public/guidelines/hhi.html.
8 Steger, 2003, "Presidential Renomination Challenges in the 20th Century."
9 Karol, 2009, *Party Position Change in American Politics*.
10 Steven J. Brams, 1978, *The Presidential Election Game*, New Haven, CT: Yale University Press.
11 Steger, 2003, "Presidential Renomination Challenges in the 20th Century."
12 Aldrich, 1980, *Before the Convention*; Bartels, 1988, *Presidential Primaries*.
13 Mayer, 1996, *The Divided Democrats*.
14 Arthur Paulson, 2007, *Electoral Realignment and the Outlook for American Democracy*, Boston: Northeastern University Press.
15 Adkins and Dowdle, 2004, "Bumps on the Road to the White House."
16 Smith and Moore, 2015, *Out of the Gate*.
17 Steger, 2013, "Two Paradigms of Presidential Nominations."

7

COLLUSION OR COMPETITION
DURING THE INVISIBLE PRIMARY

Why are the primaries more competitive in some election years than in others? It depends mainly on what happens before the caucuses and primaries begin. When presidential nominations are effectively decided *before* the caucuses and primaries, it is usually because party stakeholders and party identifiers rally around a candidate during the invisible primary. When these groups fail to coalesce behind a candidate, then there is more uncertainty about which candidate will win and voters have more viable candidates to select among in the caucuses and primaries. As we will see in the next two chapters, the conditions that contribute to party disunity also invite more candidates to enter the race since more candidates are likely to see sources of support on which they hope to build a winning coalition. Uncertainty also enables more of the candidates in the race to claim that they have a chance, raise money, and draw the media attention needed to campaign more effectively.

Whether and to what extent nominations are effectively determined during the invisible primary can be assessed by looking at insider support, campaign funding, and support for candidates in national opinion polls of party identifiers.

Competition and Insider Support During
the Invisible Primary

The process of building a winning nominating coalition begins well before the caucuses and primaries begin. Potential candidates try to get a sense of their chances of winning the nomination two or three years before the election year. Candidates try to gain the support of party insiders, leaders of groups affiliated with the political parties, party activists, and rank-and-file party identifiers. Party insiders are looking mainly for a candidate who can win the nomination and general election, and they are especially concerned with how a given candidate will play with their own constituents. They do not want a presidential candidate who will hurt the party's chances in congressional elections. Party activists and group leaders, on the other hand, evaluate the candidates to get a sense of which candidates will be the best champion of their ideological or policy cause. These participants pay attention to the candidates and to each other, trying to get a sense of which candidate is getting support from the other constituencies in the party.

Although it is an ambiguous process, it is possible to evaluate the extent of coalition-building during the invisible primary to get a sense of how much party stakeholders and party identifiers are unifying behind a given candidate or whether they are remaining uncommitted or divided among the candidates. If there is substantial stakeholder coalescence around a given candidate, then there will be a high degree of concentration of insider endorsements, fundraising, and support for candidates in national polls of party identifiers by the end of the invisible primary.

While the value and visibility of individual endorsements vary, it is the overall pattern of endorsements that provides an indication of insider support for candidates' campaign efforts, as explained in chapter three. Candidates gaining more endorsements are likely to have advantages in campaign resources and exposure and they appear to be viable and electable to those paying attention. If party stakeholders concentrate their endorsements around one or two candidates, they send a powerful signal to contributors, activists, the media, and primary voters as to which candidates should be given attention and support. If stakeholders refrain from making endorsements or divide their endorsements among a number of candidates, the signal sent to these attentive publics is weaker and less indicative of who should be supported. The race will have more competitive balance among the candidates if party insiders are divided among the candidates or if they take a wait-and-see approach.

In general, party insiders concentrate their support for candidates more than do voters in the primaries, as noted in chapter six. Figure 7.1 shows the degree of concentration of party insider support for presidential candidates in open nominations from 1976 through 2012. The HHI provides a measure of the balance among candidates in the competition for endorsements. Nominations below the .25 threshold are races with a competitive balance among the candidates, indicating that party elites were divided in their support for candidates. Nominations above that level have an imbalance of competition because party elites

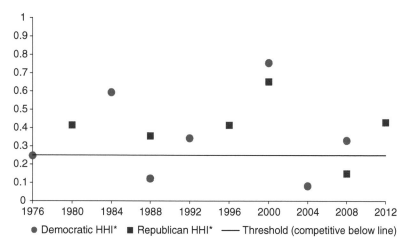

Figure 7.1 Competition for Elite Endorsements by the End of the Invisible Primary, 1976 to 2012

Source: Author's calculations using HHI*, a *normalized* Hirschman-Herfindahl Index of candidate shares of endorsements made by governors, senators, and U.S. representatives.

coalesced behind one or maybe two candidates substantially more than others. Figure 7.1 shows that most of the time party elites were relatively unified in their support for candidates, though there is a notable difference between Democrats and Republicans in this respect.

Only in four open nomination campaigns between 1976 and 2012 did party insiders fail to coalesce behind a candidate who emerged as the front-runner and went on to win the nomination (there are not equivalent data to make assessments about 1972, though it is likely that Democratic elites failed to unify behind a candidate in that year). Elite Democratic elected officials failed to unify behind a candidate in 1976, 1988, and 2004. Elite Republican elected officials failed to unify in 2008. In addition, it should be noted that the preferred candidate of Democratic insiders was rejected by caucus and primary voters in 2008. Hillary Clinton had the support of more party stakeholders than any other Democratic candidate during the invisible primary. A majority of the members of the Black Congressional Caucus supported Clinton— until Barack Obama won the Iowa caucus and demonstrated that he could win in a predominantly white state.

Democratic Party insiders coalesced around candidates in four of the seven open presidential nominations since 1976, and their preferred candidate was rejected in 2008. It should be noted, however, that many

Democratic stakeholders refrained from endorsing any candidate in that year. A lot of party insiders were hesitant to back Clinton, which gave other candidates a bigger window to advance their own ambitions. It is unclear whether Democratic stakeholders can influence the nomination even when they unify behind a front-runner. Democratic Party elites strongly unified behind Walter Mondale in 1984 yet that race remained competitive during the primaries because party identifiers were not nearly as supportive of Mondale once the voting began. Until recently, Democratic insiders were more likely than Republican insiders to *refrain* from making any endorsement or to be more *divided* among the candidates. Democratic nominations have been more competitive and uncertain at the end of the invisible primary compared to Republican nominations. With more uncertainty about who will be nominated, fewer Democratic office holders are willing to endorse any candidate before the caucuses and primaries. The competitiveness of Democratic campaigns probably makes elite office-holders more hesitant to stick their own necks out to back a candidate. The lack of endorsements and uncertainty in turn make these races more competitive by denying the media, campaign contributors, party activists, and party identifiers a clear signal about which candidate should be nominated. In open nominations in the modern era, Republican stakeholders have coalesced behind a candidate during the invisible primary in all but the 2008 nomination race when Republican insiders divided their support. Republican insiders appear to have been wary of Senator John McCain's commitment to tax cuts and his tendency to take stands against the party on campaign finance reform and other issues. Insiders are not fond of mavericks. Republican insiders were relatively unified in their support of Mitt Romney in 2012—but importantly and in contrast to prior Republican nomination cycles, most insiders refrained from making an endorsement in 2012. While Romney gathered more insider support than other Republican candidates, Republican stakeholders largely sat out the 2012 nomination race. Republican insiders remained divided or undecided about which candidate to support. It does not seem surprising then that the 2008 and 2012 Republican primaries were more competitive than most.

Important groups that are active in Republican Party politics also took a wait-and-see approach during the invisible primary leading up to 2012. In particular, many evangelical Christians were undecided and divided over the candidates. Many if not most of these groups did not endorse a candidate during the invisible primary. Only after the Iowa caucus and New Hampshire primary did a concerted effort occur to sway the outcome of the nomination. At that time, 122 leaders of various Christian groups met in Texas to endorse a Republican presidential candidate. On the first ballot, the leaders had 57 votes for Rick Santorum, 48 for Newt Gingrich, 13 for Rick Perry, 3 for Mitt Romney, and 1 for Ron Paul.[1] After two more rounds of balloting, the meeting resulted in 85 leaders endorsing Santorum and 29 endorsing Gingrich. Eight of the original supporters of Newt Gingrich walked out of the meeting once it became clear that their preferred candidate wasn't going to be supported.[2]

This case illustrates several points. One, these leaders of one of the main constituencies of the Republican Party did not endorse a candidate during the invisible primary and so could not have helped determine the nominee before the voting began. Second, the leaders of these groups met to find an acceptable alternative to Mitt Romney. The preferred choice of elite elected officials was not supported by the leaders of social conservative groups, which indicates divisions within the party coalition. Third, even when they did try to coordinate their efforts, these leaders divided among themselves. These leaders did not prefer Romney, but they could not decide on an alternative until after Rick Santorum gained momentum by narrowly beating Romney in the Iowa caucus. Such divisions can occur among the myriad of groups that form the constituencies of the political parties at the national level—even among group leaders who share a commitment to a common set of values such as social conservatism. Finally, Romney prevailed in the primaries despite a lack of support from—if not opposition by—the leaders of major groups affiliated with the Republican Party. Party stakeholders can influence and perhaps even pre-determine the nominee during the invisible primary—but only if most of them come to agreement on a candidate early enough to affect the campaign and to signal the larger population of partisans who will vote in the elections.

Relating these observations from 2008 to other primaries discussed in the previous chapter, it is apparent that the primaries are more competitive when party elites are divided or remain uncommitted during the invisible primary. When the majority of party elites and groups have rallied behind a candidate during the invisible primary, the caucuses and primaries have been less competitive. The Republican Party's primaries have been less competitive in most years because party elites have coalesced behind a candidate. When stakeholders have not unified early, the primaries have been more competitive. Democratic primaries have been competitive in most years. The main limitation on the influence of Democratic Party elites is simply that most of them remain uncommitted during the invisible primary. A majority of elite party officials refrained from making an endorsement in all but 1984 and 2000 when a former or current vice president entered the race (and the Democratic primaries of 1984 were competitive even though party elites had coalesced around Mondale). Campaigns in which a Democratic candidate secured only a plurality of elite endorsements were competitive during the primaries. On two occasions—2004 and 2008— the candidate with the most elite endorsements on the eve of the Iowa caucus did not win the nomination. In both cases, many Democratic elites took a wait-and-see approach to see which candidate emerged with the most appeal among Democrats voting in the primaries.

Once the voting begins, the results of caucuses and primaries carry more weight with the voters in subsequent contests than do the signals of support by party elites and group leaders. For example, the previously mentioned January meeting of Christian leaders was reported in news programs and talk radio programs in Southern states as well as in postings on internet blogs. But the endorsements of these group leaders received vastly less attention in the media and online than did the results of the Iowa caucus and the New Hampshire primary. The meeting of evangelical leaders in January 2012 was too late to have a substantial impact on the subsequent primaries.

When party stakeholders remain uncommitted or divided during the invisible primary, voters will be influenced more by the results of the elections occurring earlier in the sequence of caucus and primaries. As a result, candidates who "beat expectations" in a caucus or primary— especially the early ones—get more attention that is more favorable (as Santorum did when he narrowly beat Romney in the Iowa caucus, though it should be noted that the media initially reported that Romney had narrowly won). Santorum's emergence reflects an effect of momentum—he had no support from Republican elite elected officials and lagged well behind in money and polls prior to the Iowa caucus. After Iowa, he emerged as an option to Romney that was taken more seriously by some party stakeholders. Ultimately, however, Santorum could not match Romney's money and organization, he had almost no support from party insiders, and he was less appealing to large numbers of primary voters. Santorum received about 20% of the votes in the 2012 primaries and finished a distant second to Romney in convention delegates.

Thus in both political parties, party stakeholders can influence the nomination campaign but only to the extent that they are willing to unify and commit to a candidate well in advance of the caucuses and primaries. In almost every nomination campaign when the majority of party insiders have unified behind a candidate during the invisible primary, there was relatively less competition during the primaries, as indicated in Figure 6.2. Democratic primaries have been relatively more competitive, in part because Democratic insiders often refrain from making public commitments to presidential candidates or they divide their support among several candidates, such that no candidate emerges as a particularly strong front-runner. Republican insiders acted similarly in 2008 and 2012. What this suggests about the two political parties and the candidates will be discussed in the following chapters.

Competition and Campaign Fundraising During the Invisible Primary

The competitiveness of presidential nomination campaigns also can be evaluated by looking at the concentration of campaign funds. The rising costs of candidate-centered campaigns together with a compressed primary schedule have produced a major competitive gap between candidates who can obtain resources and those who cannot.[3] The rising costs of campaigns mean that a candidate's ability to compete depends heavily on their ability to raise money. Well-financed candidates can build more extensive organizations at the national and state levels; they can afford the market research needed to craft messages with the desired appeal and to target their resources more effectively; and they will be more capable of identifying, communicating with, and mobilizing potential supporters. Well-funded candidates appear to be viable contenders.

Candidates' relative abilities to compete *during* the primaries have been heavily affected by the pre-primary competition for resources. As explained in chapter four, money has the greatest impact on political campaigns when there is a substantial imbalance of resources across the candidates. When one candidate has most of the resources, that candidate has greater capacity to communicate with voters. If one candidate raises substantially more funds than other candidates, then the nomination race could be decided before the caucuses and primaries through the collective decisions of party activists who donate to presidential nomination campaigns.

The money chase among candidates is usually competitive during the invisible primary (see Figure 7.2). There is more balance in the distribution of funds across candidates by the end of the invisible primary compared to the competitive balance of endorsements or of voting

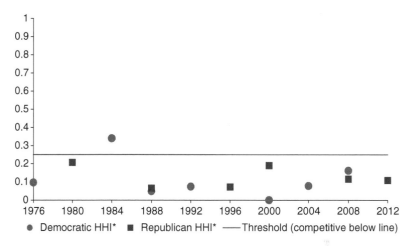

Figure 7.2 Competition for Funds by the End of the Invisible Primary, 1976 to 2012

Source: Author's calculations using HHI*, a *normalized* Hirschman-Herfindahl Index of candidate shares of campaign funds during the invisible primary.

in the primaries. This is a major reason that fundraising has not been found to be a significant predictor of which candidate will win a presidential nomination. The leading fundraiser rarely has a big enough lead in money to eliminate other candidates. This makes presidential nominations quite different from congressional elections, where there usually is one candidate with a big enough advantage in funds that these elections often are not competitive.

This does not mean that money does not matter. Candidates need money to compete. What the evidence shows is that *money raised during the invisible primary does not determine the winner in presidential nominations.* There usually are several candidates who can raise enough money to make their case to voters, so money does not deny caucus and primary voters a meaningful choice of candidates in presidential nominations. In only one nomination race was there enough of an imbalance in campaign funds to say that the race was not competitive on this dimension. Walter Mondale had a substantial advantage in campaign funds in 1984. That also was a year in which Democratic elites had unified behind Mondale's candidacy. Despite these advantages, the 1984 primaries were competitive as Gary Hart rose out of nowhere to wage a strong campaign after gaining momentum in the Iowa caucus and New Hampshire primary. Harts' momentum, however, was not sufficient to overcome Mondale's advantages in insider support and campaign funds. This case suggests that it is possible that money can help make a difference in determining the outcome, but only if campaign contributors concentrate their donations to one candidate. In most presidential nominations, they do not. This is a substantial difference between campaigns at the presidential and congressional levels. Campaigns for Congress usually do have highly concentrated fundraising patterns so that money does impact who wins.

Competition and Support Among Party Identifiers During the Invisible Primary

The success of candidates ultimately depends on their abilities to appeal for the support of large numbers of party activists and identifiers. Compared to party insiders, party identifiers tend to be more divided in their support for presidential candidates. While insiders usually coalesce behind one or two candidates, party identifiers more often divided during the invisible primary in most years. Democratic Party identifiers failed to give a majority of their support to a candidate in all but 1984 when Walter Mondale led in national public opinion polls during the invisible primary. Even in 2000, Democratic Party identifiers were almost evenly divided between Vice President Al Gore and former Senator Bill Bradley. Democratic nomination races usually have had a competitive balance among the candidates when it comes to the support of party identifiers. They rarely have a front-runner who has the support of a majority of respondents in national opinion polls. Not surprisingly, Democratic primaries are usually competitive with several candidates attracting support deeper into the primary season than is the case for Republicans.

Republican Party identifiers were somewhat more unified in their support of candidates, particularly in 1980 when Reagan was quite popular among Republicans and in 2000 when George W. Bush had majority support in national public opinion polls. Republicans were generally unified behind Vice President George H. W. Bush in 1988 and Senate Majority leader Bob Dole in 1996. Republican identifiers were more divided during the invisible primaries for the 2008 and 2012 nominations. These two nominations were also relatively competitive into the caucus and primary season.

For both party insiders and party identifiers, there is evidence that the political parties unify behind a front-runner more in some years than in other years. Both insiders and party identifiers were unified to a considerable extent behind Ronald Reagan, George H. W. Bush,

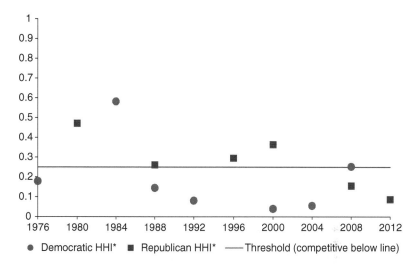

Figure 7.3 Competition for Public Support by the End of the Invisible Primary, 1976 to 2012

Source: Author's calculations using HHI*, a *normalized* Hirschman-Herfindahl Index of candidate shares of support among party identifiers and independent leaners in national Gallup polls during the fourth quarter of the year before the election.

Bob Dole, and George W. Bush in their presidential nomination campaigns. All of these candidates went on to win the Republican nominations despite losing one of the early caucuses or primaries. What these differences between the political parties tell us will be the subject of the next chapters.

Conclusions

Analyses of the factors during the invisible primaries of open presidential nominations look different from the picture provided by an analysis of competition in the presidential primaries. Democratic elites were relatively divided in 1976, 1988, and 2004 and somewhat more unified in other years, though many Democratic elites remained uncommitted in 1992 and 2008. Republican elites were divided only in 2008 and most remained uncommitted during the invisible primary in 2012. Looking at the competitiveness of the primaries and the cases in which elites unified, the invisible primary phase of the campaign has been more determinative of the outcome for Republicans than Democrats. When Republican elites unified, their preferred choice won and there was relatively little competition in the primaries. Republicans have generally unified behind a front-runner during the invisible primary, breaking from that pattern only in 2008 and 2012—primaries which were more competitive than most.

Democratic elites have been more likely to take a wait-and-see approach or to divide their support, and Democratic primaries have been more competitive. The main exceptions to this occurred in 1984 and 2000 when Democratic insiders unified behind a vice president who was seeking to become the party's nominee. The Democratic primaries were competitive in one (1984) of these two races. The reverse happened in 2004 when Democratic elites were divided among the candidates while voters in the Democratic primaries quickly coalesced around John Kerry's candidacy. While there are few cases on which to base generalizations, it does not seem to make a great deal of difference in Democratic races whether elites unify or not when it comes to the competitiveness of the primaries. A major reason may be that Democratic elites tend to remain uncommitted during the invisible primary.

There is little evidence that campaign donors are able to coordinate among themselves enough to give one candidate enough of an advantage to ensure a victory. While not all party activists donate to a candidate, virtually all donors are party activists. In this respect, campaign contributions can be viewed as an indicator of support for candidates among party activists. If this assumption is accepted, then it would appear that party activists are relatively divided in their financial support for candidates. In almost every nomination campaign, there are several candidates who are able to raise enough money to compete at least through the early contests. Collusion by campaign contributors during the invisible primary phase of the campaign does not pre-ordain the outcome. Contributions during the invisible primary also are less of a factor now that candidates can raise large sums of money online from party activists or from billionaires who can single-handedly give a candidate enough money to compete in particular primaries. If anything, money has become less determinative of the outcome in recent nominations than it was previously when candidates had to develop large networks of donors on a national scale.

Polls of party identifiers and independent voters who lean toward a political party give us an indication of candidate support among the mass membership of a political party. By this measure, all of the Democratic nomination races have been at least moderately competitive during the invisible primaries from 1972 to 2012, with the sole exception of 1984 when Mondale was a dominant candidate until the primaries.

The mass membership of the Republican Party has unified behind a front-runner during the invisible primary in every year except 2008 and 2012. There is much more alignment between the candidate preferences of Republican Party elites and those of rank-and-file Republicans. There is less of a correspondence between the candidate preferences of Democratic elites and those of Democratic identifiers.

It is not a coincidence that Democratic primaries have been more competitive than Republican primaries—at least until the last two elections when Republicans showed more signs of intra-party divisions at the elite and mass membership levels. While Democratic elites have been able to unify in some years, their influence on the mass membership of the Democratic Party seems to be more tenuous. Campaign momentum likely played a big role in the nominations of Democrats Jimmy Carter, Michael Dukakis, Bill Clinton, John Kerry, and Barack Obama. Only Republicans John McCain and Mitt Romney could be argued to have been nominated as a result of gaining campaign momentum during the primaries. In Romney's case, campaign momentum likely helped to firm up support among those Republicans who were wary of him.

The question we will now turn to is, "Why do party stakeholders and party identifiers coalesce behind a front-runner during the invisible primary of some nominations more than others?" The focus will be on these themes—the relative unity of the party coalitions and the decisions of potential candidates to enter the race.

Notes

1 Paul Stanley, 2012, "Gingrich Camp Cries Foul Over Evangelical Leaders Endorsement of Santorum," *Christian Post*, January 17, www.christianpost.com/news/gingrich-camp-cries-foul-over-evangelical-leaders-endorsement-of-santorum-67393/.

2 Erik Eckholm, 2012, "Four Evangelical Leaders Reaffirm Support for Gingrich," *New York Times*, January 16; Stanley, 2012, "Gingrich Camp Cries Foul."

3 Randall E. Adkins and Andrew J. Dowdle, 2002, "The Money Primary: What Influences the Outcome of Presidential Nomination Fundraising?" *Presidential Studies Quarterly* 32(2): 256–275.

Part III

EXPLAINING COALITION FORMATION IN PRESIDENTIAL NOMINATIONS

8

POLITICAL PARTY UNITY
Long- and Short-Term Variations

The argument that the invisible primary is the critical period of the nomination campaign is contingent on the unity of the party coalition. The extent to which parties will coalesce around a candidate during the invisible primary depends in part on the relative unity of the party coalition.[1] While the coalitions that form the parties are durable, some degree of coalition change occurs in every presidential nomination.[2] Both long- and short-term factors affect party unity in presidential nomination campaigns.

The coalitions of the major parties have undergone several transformations of their membership throughout American history. These party realignments involve long-lasting—but not necessarily permanent—changes in the membership of the party coalitions. These changes occur for various reasons such as a response to cataclysmic events like the great depression; new groups entering the electorate through generational change and immigration; and a response to issue evolutions as party leaders seek to use issues that cut across existing party cleavages.[3] Presidential candidates and presidents also contribute to patterns of party coalition formation, maintenance, and fragmentation.[4] The past fifty years have witnessed a party realignment as citizens "sorted" into parties as the ideological reputations of the parties have become clearer to citizens.[5] Whatever the cause, these long-term changes in the party coalitions affect the constituencies that participate in each party's nomination of presidential candidates.

There also are short-term changes in the coalitions that nominate presidents. While political parties are a broad collection of constituencies, not all of the citizens and groups that affiliate with a party will participate in the selection process in a given nomination cycle. As discussed in chapter two, the main consequence of the 1970s era reforms was to formalize the participation of party activists in the selection of presidential nominees. The set of party activists and party identifiers who participate in any given presidential nomination varies somewhat from election to election. A presidential nominating coalition must be assembled or reassembled every four years. This means that the constituencies that nominate a president can change more rapidly than does the broader membership of a party coalition.

The ability of party stakeholders to coalesce behind a candidate during the invisible primary is affected by variations in the unity and stability of the party coalition. The unity and stability of party coalitions varies with long-term changes in the citizens and groups that form the party coalitions and to a lesser extent by short-term differences in which citizens and groups participate in nomination campaigns.

Changing Political Party Coalitions

The 1960s through the early 1990s was a period of instability and transition in both the Democratic and the Republican Party coalitions. One major consequence was that the presidential nomination campaigns of both political parties were more open and competitive compared to nominations occurring in more recent years. The Republican Party gained a greater degree of unity by the 1980s than did the Democratic Party, which has continued to exhibit greater internal divisions.[6] This relatively greater unity helps explain why Republicans were better able to coalesce around presidential candidates during the invisible primaries from 1980 to 2008 compared to the Democrats.

The last fifty years have witnessed tremendous economic, demographic, social, technological, and political changes in the United States. The country's economy has transformed from industrial manufacturing to an economy anchored in the technology and service sectors. The computer and internet revolutions propelled massive gains in productivity which accompanied a massive growth of wealth at the top with wage stagnation for the middle and lower segments of the population. Growing socioeconomic inequality has dramatically affected the political parties, contributing to an increasingly liberal Democratic Party and a more conservative Republican Party.[7]

Demographically, the United States is becoming a more brown country as immigration shifted from Europe to Asia and South America. Higher divorce rates and more children born out of wedlock mean more single-parent households, which correlates strongly with poverty, criminal activity, and low performance in education, which all limit upward mobility. An increasing percentage of two-parent households have both parents working outside the household. An aging population threatens to sap a social welfare system in which most benefits go from working adults to retired adults, and in which Medicare expenses rise as an older population requires more medical care. Socially, urban cities have lost population since 1970 and about half of Americans live in suburban areas. The country has witnessed liberal and conservative social movements advancing the civil rights of African Americans, Latinos, women, and GLBT populations on the left and more religiously motivated traditional values on the right. Technological changes in communication and the internet continue to transform the economy, social interactions, and the nation's culture. All of these changes have affected politics in the United States, especially our long-term political coalitions.

There was considerable instability in the electoral coalitions of both parties from the 1960s through the 1970s.[8] These years featured a substantial change in the electoral bases of both parties, at least at the national level in presidential elections. The realignment of the political parties continued through the 1980s into the 1990s for elections to lower level offices. These changes are socioeconomic, cultural, demographic, and geographic in scope. Socioeconomically, wealthier Americans became less likely to identify with the Democratic Party and more likely to support the Republicans. Republican gains in the South, for example, have been strongest among more affluent, white voters.[9] In part this owes to generational change—the conservative Democrats of Southern states during the 1950s and 1960s have been replaced by conservative Southerners who identify with the Republican Party. Similarly, lower-middle income voters became less likely to support Republicans and more likely to identify with the Democratic Party. The famous "Reagan Democrats" who supported Reagan in 1980 were in part the rank-and-file of organized labor, but in most cases these were blue-collar workers and tradesmen who had middle- or upper-middle class incomes. Citizens with lower incomes are more likely to identify with the Democratic Party today than they were in the 1950s.

Culturally, the rise of social and cultural issues contributed to changing coalitions of the two political parties. These issues are not unrelated to changing socioeconomic conditions. For example, the tremendous increase of women in the workforce has contributed to a split along social or cultural divisions. On the left, women sought greater equality in pay and career options. The rise of women as an independent voice in society is threatening to those whose beliefs emphasize traditional gender roles with women as homemakers in a nuclear family with the father as the provider. Issues like abortion are not just about moral values; the issue also relates to conflict over broader conceptions of the role of women in society. Perhaps not surprisingly, women tend to divide their allegiances between the two political parties. Single women are more likely to identify as Democrats while married women are more likely to identify as Republicans.

The country's liberal and conservative movements have had substantial effects on the activist base of each major party. Since the 1960s, the Democratic Party has absorbed substantial increases in the participation of liberal activists seeking to advance the civil rights of African Americans, Latinos, women, and GLBT populations, along with a vocal environmental movement. The Democratic Party has experienced a decline in its active membership among Southern whites and traditional labor unions. While Southern whites and labor unions still matter, there is little doubt that unions and white Southern Democrats have less influence in the Democratic Party than they did during the 1940s to the 1970s. Off-setting these losses have been increases in identification with the Democratic Party in Northern and West coast states.

The Republican Party has experienced similar change in its activist base. Most notably, religiously motivated proponents of traditional values have raised the salience of civil liberties—issues relating to church-state relations like abortion and gay marriage, gun control, and issues relating to criminal or victims' rights (depending on one's preferred framing of these issues). The rise of cultural conservatism relates to socioeconomic status. Although a few political commentators have argued that Republican strategists use social issues to attract support of lower income voters, it is more affluent, white evangelical and Pentecostal Christians who are more likely to self-identify as social conservatives who support the Republican Party. Thus allegiance to the Republican Party is grounded in both economic and social conservatism.

The traditional, more "libertarian" wing of the Republican Party has seen its voice decline in presidential nominations. Republicans with a libertarian orientation are economically conservative—wanting less government regulation and lower taxes like other conservatives—but they are moderate to liberal on social issues. Socially moderate candidates have not done well in Republican presidential nominations since Ronald Reagan became president. The last major socially moderate Republican presidential candidates—Pennsylvania Senator Arlen Specter and California Governor Pete Wilson—dropped out of the race even before the caucuses and primaries began in 1996. Recent libertarian-type Republican candidates like Ron Paul have received only minor support.

These changes correspond to uneven economic growth across states. Geographically, states that experienced declining real median household income have become more Democratic, while states that have had gains in household income have become more Republican. Republican Party identification has grown in Southern and Southwestern states that have seen the greatest gains in household income. Democratic Party identification has grown in Northern states whose economies suffered from the declining employment and wages in manufacturing.

These transformations of the electoral bases of the two political parties have changed the kinds of candidates and policies that are advanced during presidential nominations and the general election. The reforms of the 1970s enabled political activists of liberal and conservative movements to advance policy change through the Democratic and Republican Parties, respectively. Nominations are the vehicle that activists use to promote their policy goals. The result is a transformed nomination process with political parties that are substantially different from those of the earlier half of the 20th century. Party insiders and candidates cannot control party activists as party bosses could do with their patronage workers in the 19th and early 20th centuries.[10] Elected officials are the product of these same forces, so to a considerable extent, party insiders have greater alignment in their political and policy preferences with the activist bases of their parties today than they may have had in the past.

Periods of change in the party coalitions relate to the competitiveness and outcomes of presidential nomination campaigns. Although the process of partisan realignment began in the 1960s and continued through the 1980s, the 1970s witnessed the greatest degree of change in the electoral bases of the political parties in presidential politics. This period of transition coincides with the presidential nominations that had the greatest degree of competition among candidates before and during the primaries.

With so much of the membership in flux, there was considerable uncertainty about which candidates could win in both political parties. This kind of uncertainty itself is a reason for more candidates to take their chances and enter the race. That calculus and the entrance of more major candidates vying for a nomination adds to the challenges that party stakeholders face when seeking to coalesce around a given candidate during the invisible primary. It also makes it more likely that the nominations would be decided during the primaries.

The nomination and election of Ronald Reagan seems to have settled or at least reduced the intra-party conflict on the Republican side.

From 1980 until 2008, the Republican nominations exhibited less competition, less uncertainty, and tended to be settled before the caucuses and primaries began. This coincides with a period of greater coalitional stability in the Republican Party. That the Republican nominations of 2008 and 2012 appear to be different could indicate that this peace within the Republican Party coalition may be coming to an end. The divisions that appear to be emerging in the Republican Party have several facets. There are conflicts within the party between libertarians and social conservatives, though social conservatives continue to hold sway for now. There also is a division between Republicans who favor a more isolationist "America First" foreign policy and a more dominant Republican sentiment that favors a "strong national defense." More pressing, however, seem to be divisions between an emergent economic populist movement in the Republican Party and the more traditional pro-business faction of the party. The Tea Party movement is itself a diverse collection of groups. However, part of the movement involves economically conservative populists who opposed the government bailouts of banks and other industries in 2008 and 2009. These conservatives see Republican support for bailing out Wall Street as abandoning the conservative commitment to free-market principles, which in their pure form require that companies be allowed to fail. While many economists believe allowing major financial institutions to collapse would have had devastating economic effects, many economically conservative populists remain unconvinced.

While none of these disagreements has torn apart the Republican Party, these internal divisions have made the party more diverse, which in turn makes it more difficult to find a presidential candidate who appeals to all of the various groups in the party. The greater complexity of the Republican coalition makes it less likely that the party will unify behind a candidate during the invisible primary and more likely that presidential nominations will be contested through the caucuses and primaries in which one candidate will seal the deal by gaining momentum across these contests.

There is a similar pattern of transition in the Democratic Party coalition, although it is less clear that the Democratic Party attained the same kind of coalitional stability that the Republicans did during and after the Reagan years. The Democratic Party coalition that emerged in the 1930s during the "New Deal" of Franklin D. Roosevelt fragmented beginning in 1938 and increasingly so by the 1960s. The Democratic Party has remained relatively divided since then.[11] Social conservative candidates quit trying to gain the Democratic presidential nomination.

The last overtly pro-life candidate to seek the Democratic presidential nomination was a legal secretary from New York named Ellen McCormack in 1976. The high point of her campaign occurred when she won 14% of the vote in the 1976 Kentucky primary—an indication of the power of this issue among evangelical Christian voters. Other Democratic candidates switched from a pro-life position to pro-choice position prior to (and in anticipation of) their campaign. For example, Senator Ted Kennedy took pro-life positions in the 1970s, switching to a pro-choice position in 1979 as he geared up to challenge Jimmy Carter's renomination in 1980. Vice President Al Gore switched his position from pro-life to pro-choice before running for the 1988 Democratic nomination. Gore had sponsored one of the first bills to restrict abortion in 1975 when he was still a junior U.S. senator. More recently, U.S. Representative Dennis Kucinich switched his position from pro-life to pro-choice before seeking the Democratic nomination in 2004. Both Republican and Democratic presidential candidates now adhere to the position that became party orthodoxy on social issues like abortion. The main divisions in each party are over the relative primacy of various economic policies.

The Democratic Party coalition is, in many respects, more complex than that of the Republican Party. The Democratic coalition is a blend of economic liberals, social liberals, labor groups, environmental activists, urban voters, and a variety of ethnic and racial minorities. Some of these groups have clashed with more moderate Democrats who organized in the Democratic Leadership Council, a group that more openly embraces the free-market and supported policies like welfare reform in the 1990s. These moderate Democrats sought to expand the party's appeal among more affluent voters, particularly in the suburbs where about half of all Americans live. The modern Democratic Party coalition also continues to shift with the changing demographics of the United States, changing in particular as more of the growing Hispanic voting population identifies with the Democratic Party. This creates some tension with African American voters, who tend to be less supportive of liberalizing immigration laws. In a coalition as diverse as the Democratic Party, coming to agreement on a presidential candidate is challenging.

It should be noted that neither party is completely unified on any issue. For example, there are social conservatives and social liberals in both political parties at the congressional level.[12] One reason is that congressional candidates deal with much less diverse populations and constituencies. For example, there is a tendency for Democrats representing rural and Southern areas to be pro-life, and for Republicans representing urban areas in Northern states to be pro-choice. Congressional candidates of both political parties adopt the majority position of the local population on social issues. Thus congressional Democrats from Kansas or Oklahoma tend to be pro-life and congressional Republicans from New England tend to be pro-choice. While there may be more homogeneity of preferences at the local level, there is a lot more diversity of policy preferences at the national or presidential level of party nominations.

At the presidential level, candidates tend to adopt positions on high salience issues that are held by the majority of political activists in the party. A candidate who does not adapt to the prevailing positions on salient issues will not likely have a realistic chance of winning the presidential nomination of that party. Party positions on issues evolve and change as do the coalitions of the parties. These changes create potential for conflict and competition within the party over the nomination of a presidential candidate.

Eras of relatively high coalitional instability—as occurred in the 1960s and the 1970s—tend to be associated with greater competition among candidates and less cohesion among party insiders and activists during the invisible primaries, as we saw in the last chapter. Democrats tended to have less unity behind their presidential candidates during the invisible primaries, unifying behind a candidate only in 1984 and 2000 when the front-runner was a vice president. Republicans were highly divided in 1976, but unified substantially during the Reagan years and continuing through the George W. Bush Administration. Republican Party stakeholders and primary voters have been more divided in the last two nominations. The extent to which the parties have coalesced behind a presidential candidate during the invisible primary depends a lot on the unity of the party coalitions.

Short-Term Forces and Participation in Presidential Nominations

There also are short-term forces that create variations in the coalitions that nominate presidential candidates. The major American political parties are broad coalitions of different groups and constituencies. In any broad collective of people and groups, there are multiple possible winning coalitions of members within the broader collective. While the party coalitions change slowly and are similar from election to election, the party constituencies that join together in a winning *nominating coalition* are reconfigured on a national scale every four years in a presidential election. The coalitions that nominate presidential candidates can change from election to election because of variation in participation by party constituencies. Variations in patterns of participation in the nomination process are intertwined with whether and to what extent party stakeholders coalesce behind a candidate during the invisible primary. We have seen that some party stakeholders make endorsements in some nomination campaigns while taking a wait-and-see approach in others. The variation in participation by elite office holders is especially stark. For example, 15% of Democratic governors endorsed a candidate during the invisible primary of 1988, while 59% endorsed a candidate during the invisible primary in 2000. The percentage of Democratic senators endorsing a candidate during the invisible primary has varied from a low of 14% in 2004 to 59% in 2000. On the Republican side, all of the governors endorsed a candidate—mainly George W. Bush— during the invisible primary leading up to the 2000 Republican nomination while only 36% of the governors endorsed a candidate during the invisible primary in 2008. It is hard to argue that party insiders play a decisive role in the nomination when a majority of them take a wait-and-see approach to the nomination race. Nominations in which party insiders remain uncommitted are more likely to be decided during the caucuses and primaries.

Voter turnout in presidential caucuses and primaries also varies significantly from election year to election year. For example, 14 million citizens voted in the 2000 Democratic primaries while more than 37.2 million citizens voted in the 2008 Democratic primaries. Turnout in Republican primaries has varied from a low of 12.2 million voting in 1988 to 20.6 million in 2008. Variation in participation across election years means that there are differences in the coalitions that nominate candidates in each of the political parties. Interestingly, there is no significant correlation between voter turnout in the primaries and the proportion of party elites making an endorsement during the invisible primary. It is not the case that voter turnout is lower in nomination races in which party insiders unify behind a candidate during the invisible primary. Turnout in the primaries seems to be more a function of the number of viable candidates throughout the primaries, which is only partially affected by party insider activity.

Party groups and constituencies vary in their involvement in a given presidential nomination campaign. Participation varies in part because of the costs and benefits of political activism. Engaging in the nomination is costly in terms of the time spent learning about candidates, encouraging associates to support a candidate, volunteering, contributing money, and so on. There are opportunity costs since time spent on political activities is not available for other pursuits. As a result, sustaining political activism across time is hard. Groups and citizens may participate in some nominations but not others. While the costs of participation are fairly constant, the benefits are more variable so understanding why stakeholders and party identifiers get more involved in certain nominations requires looking at the benefits of involvement. What motivates stakeholders and activists to get involved in some nominations more than others?

One major factor affecting participation is the issue environment—the set of issues that are receiving widespread attention and concern among party constituencies. Activists may be motivated to participate in a presidential nomination because they care intensely about an issue. If that issue rises to the top of the national agenda—driven largely by events—then those citizens and groups may become more active than in years when the issue is less salient. For example, the 2004 Democratic nomination was influenced in part by the participation of Democratic activists who were strongly opposed to the Iraq War. Many of these anti-war activists were Democrats who had been actively opposing the Vietnam War but apparently were not involved in nominations in the interim. Indeed, many of these middle-aged anti-war activists were first-time participants in the 2004 nomination process.[13] Their mobilization in 2004 crystalized early behind the candidacy of Howard Dean, who advocated against the war. The potency of the issue contributed to other Democratic candidates adopting a stronger anti-war position. As they did, Dean's advantage in the nomination race dwindled. Many anti-war activists saw John Kerry's experience as a Vietnam War veteran as making him a more credible candidate to lead the fight against Bush in 2004. The rise and fall of issues on the agenda affects who participates and, in turn, which candidates will be advantaged and disadvantaged by the resulting fluctuations in the nominating coalitions.

A changing issue environment also can affect the unity of a party coalition.[14] The emergence of new issues can cut across constituencies of a party and thus create opportunities for a candidate seeking to advance his or her chances of nomination. The Vietnam War certainly divided the Democratic Party in 1968, when Vice President Hubert Humphrey prevailed over anti-war candidates at the convention, as noted in chapter two. Ronald Reagan's Republican nomination in 1980 was helped by the emergence of abortion as an issue in the 1970s, leading social conservatives to join the Republican Party in large numbers. Candidates vary in their appeal on different issues and a given candidate might be a strong contender in one election but not in another, depending on which issues are salient. One of the main differences between Barack Obama and Hillary Clinton in 2008 was their positions on the Iraq War. Obama had more credibility with Democratic Party activists opposed to the war since he had opposed it from the beginning. Hillary Clinton initially supported the decision to authorize President Bush to go to war and later changed to opposing the war. Obama may not have won the party's nomination if the war had not galvanized these party activists.

Candidates are another source of short-term variation in participation in the nomination process. The presence of a particularly appealing candidate in the race can spur some party members to activism who might otherwise sit on the sidelines. Barack Obama's candidacy in 2008, for example, motivated young adults to participate in Democratic caucuses and primaries more than had happened in prior nomination campaigns. The surge of young voters participating in the Iowa caucus in January 2008 was a major source of his support in that state. Without an influx of new participants, Obama may not have gained as much of the momentum that helped propel him to the presidency. It is hard to motivate citizens to participate when the candidates in the race are not appealing. As noted earlier, evangelical Christians were not particularly enthusiastic about Mitt Romney as a Republican candidate. Fewer party elites made endorsements and fewer Republicans voted in the primaries of 2012 than had in prior nomination campaigns. So even though the Democratic and Republican Party coalitions were largely the same from 2004 to 2012, there is some variation in the involvement of particular party constituencies that came together to nominate their presidential candidates.

The presence of an obviously strong candidate especially has an effect on the involvement of party stakeholders in coalescing behind a candidate during the invisible primary. Democratic and Republican Party insiders and groups rallied very early in the invisible primary to support Al Gore and George W. Bush, respectively. In campaigns like 1992 when the early party favorite—Mario Cuomo—decided not to run, there was very little coalescing among Democratic Party stakeholders. The vast majority of party elites had sat on the sidelines waiting for Cuomo to enter the race. When he declined to run in November 1991, there was little time remaining to rally behind one of the candidates in the race. In other races, there simply is not one candidate who is clearly stronger than the others. In these races, party stakeholders often take a wait-and-see approach. By doing so, they effectively move the locus of decision to the caucuses and primaries where voters will play a more independent role in selecting the nominee.

Technological innovations in communications and market research also have affected the formation of winning nominating coalitions. Campaign organizations have gained much more ability to identify and communicate with potential voters. These advances have affected who participates in presidential caucuses and primaries as a function of campaign resources and candidate appeal. Barack Obama, for example, succeeded in increasing voter participation among young people in the 2008 Iowa caucus, an expansion of the electorate that helped him win the state and gain momentum going into subsequent primaries. More generally, Obama's use of the internet and social media is widely credited for his campaign's success in organizing and mobilizing individual supporters in the 2008 caucus states.[15] While Obama narrowly lost to Hillary Clinton in the total number of votes in Democratic primaries in 2008, he won by a wide margin in the caucus states. That enabled him to amass more delegates to the national convention. The Obama campaign's success at identifying prospective supporters and getting them to vote may have been another critical difference between winning and losing in 2008.

It should be noted that variation in participation means that there is almost always some uncertainty about who will come together to select the eventual nominees. This creates some leeway for candidates to attempt to pull together different constituencies within a party in order to form a winning coalition. Some presidential candidates assume the role of political entrepreneurs who have ideas about public policy and who want to reshape their party as well as change public policy. Such candidates may seek to change who participates in the selection of a party's nominees. Ronald Reagan's 1980 campaign, for example, sought successfully to mobilize evangelical Christians to support his nomination. Jesse Jackson's 1988 campaign sought to mobilize African Americans and other groups in the "Rainbow Coalition" in order to expand the voice of these constituencies in the Democratic Party and by doing so improve his chances of winning the nomination. There are candidates in almost every nomination cycle that seek to mobilize certain groups to gain an advantage in the caucuses and primaries.

Efforts by political entrepreneurs are resisted by party stakeholders who seek to protect party commitments to policies on issues important to them. As a result, there can be rivalry within a party as different groups and activists seek to get their preferred candidate nominated as the party's candidate. For example, supporters of fiscal conservative Republicans like George H. W. Bush, Bob Dole, or John McCain have clashed with supporters of supply-side Republicans like Ronald Reagan, Jack Kemp, and George W. Bush, respectively. While all of these Republicans would prefer balanced budgets and lower taxes, there often are disputes over priorities when a tax cut increases the budget deficit. Supply-side Republicans, for example, were skeptical of the commitment to tax cuts by Bush, Dole, and McCain. For their part, Democrats have had clashes between progressives and moderate Democrats who organized in the Democratic Leadership Council to advance more fiscally conservative policies. These Democrats supported Bill Clinton as a New Democrat who could win elections in an era when public mood had shifted in a more conservative direction.

While the existing groups and activists of a party seek to defend party orthodoxy, they may not be able to prevent change from occurring. Fiscal conservatives could not or did not prevent the nomination of Ronald Reagan who championed a version of supply-side economic theory that George H. W. Bush chided as "voodoo economics" in 1980. Supply-siders did not or perhaps could not stop the nominations of fiscal conservative Republican nominees George H. W. Bush, Bob Dole, or John McCain—though all of these candidates had to promise fealty to the goal of cutting taxes as part of their efforts to get nominated. Of these three fiscal conservatives, Bush was elected president and proceeded to compromise with Democrats in Congress to raise income taxes to help balance the budget in 1990—an example of fiscal conservatism being given priority over tax cuts. The reversal of his position on taxes was a major factor contributing to Pat Buchanan's efforts to challenge Bush's renomination for the Republican presidential nomination in 1992. The national party coalitions are not homogenous, and subtle differences in policy positions are at stake in the competition for leadership of the parties and the nation.

The 2012 Republican nomination shows that a major faction of a party may not be able to exert an effective veto over the selection of a presidential candidate that they are wary of or consider unacceptable. Many evangelical Christian leaders were not keen on nominating Mitt Romney in 2012. Romney was a Mormon, which many evangelicals do not consider to be a Christian religion. As a candidate for the U.S. senate in the 1990s, Romney had pledged to support gay rights in a letter to the Log Cabin Republicans. As governor of Massachusetts, Romney had enacted a health care law that was quite similar to the Affordable Care Act that Tea Party Republicans opposed vigorously. As noted in chapter seven, many of these leaders did not weigh in on the candidates during the invisible primary leading up to the 2012 caucuses and primaries. Yet opposition to Romney faded during the caucuses and primaries when Republican activists faced a choice between Romney and other candidates who had their own limitations. Party stakeholders may view a particular candidate as less desirable than some mythical ideal candidate, but they have to choose among the candidates who decide to run, as we will see in chapter nine. The lack of an appealing candidate may contribute to less active involvement in the nomination campaign by members of the party coalition.

The majority of people who vote or who can be prodded to vote in a caucus or primary are party activists or party identifiers who are inclined to accept candidates who preach party orthodoxy.[16] The national parties are coalitions of diverse groups and individuals with interests and preferences across a wide range of issues. While candidates may adopt the same position on a few issues that are important to most or even all of the party membership, there are a lot of issues that particular constituencies may care more about. Candidates try to appeal to targeted constituencies by raising the salience of certain issues, emphasizing certain candidate characteristics, and framing political discourse in ways that plays to their advantage. Candidates try to take advantage of these possibilities in ways that will help them get nominated.

Conclusions

It is easier to coalesce around a candidate when a party coalition is relatively stable and unified. Coalition formation tends to be *group centric* during periods of coalitional stability and especially when parties are polarized (homogenous internally and divided across the parties). Under these conditions, the various constituencies of a party are largely set and these constituencies seek a champion of their causes. Candidates compete mainly to be the champion of these constituencies. A few candidates may seek to disrupt the coalition of the party to advance their own agenda or to further their own ambitions, but these candidates will attract relatively little support. Nominations have increasingly become group centric as the political parties have become increasingly polarized along ideological lines since the 1980s.

It is harder to hold a party coalition together when a party membership is changing or when a party is internally divided among factions that may not share the same policy priorities. Individual candidates can make a difference when the parties were less well sorted along ideological lines—as was the case in the 1970s for both parties and continued to be for Democrats at least into the 1990s.[17] Upon being criticized for dividing the Democratic Party in 1968, Senator Eugene McCarthy replied, "Have you ever tried to split saw dust?" The relative lack of cohesion among Democratic Party constituencies has meant that Democratic presidential nominations often have been more competitive and uncertain during the invisible primary. More of these nominations have been decided by campaign momentum gained during the caucuses and primaries than have been decided by coalescence around a candidate during the invisible primary. Even though the Republican Party is relatively more unified than the Democratic Party, the Republican coalition has its own internal divisions. Internal divisions in a political party make it harder to find a candidate whose appeal unifies the various factions of a party. Such intra-party divisions often are episodic in that they may emerge in one nomination but not another, depending on the saliency of issues in an election year.

Whether intra-party divisions affect the nomination campaign depends in part on the extent to which underlying divisions are exposed by the issues that are salient in a given year or by candidates in a presidential nomination campaign. Candidates seek to use issues that they believe will increase their own support relative to others. These sources of coalitional variation give candidates an opportunity to develop strategy and campaign messages that will attract a winning coalition of party constituencies during the nomination campaign. A candidate must gain the support of a winning coalition of the active party membership in caucuses and primaries to become the nominee. These winning coalitions have to be reassembled every four years, leading to potential changes in the active membership and policy orientation of the parties. Ronald Reagan's success in bringing white evangelical Christians to the Republican Party speaks to the potential of a candidate to change the party coalitions. The ability of a candidate to do so depends in large part on the stability of the coalitions. It is not an accident that Reagan's success in bringing in evangelical Christians occurred at a time when the party coalitions were in the midst of a long-term realignment. As the realigned party coalitions stabilized, the potential for such political entrepreneurship likely declined until the Republican coalition forged during the Reagan years began to fragment.

In sum, the coalition-building process varies across elections even for a given political party. Changes in rules that expand or restrict participation can affect who gets involved in selecting a presidential candidate and the kinds of candidates who will have a chance of winning the nomination. The long-term changes associated with political party realignment made it harder to unify behind a candidate, particularly in the 1960s and 1970s when the party coalitions were in a period of transformation. There also are variations in the participation of party members in the selection of a nominee. These changes relate to issues, candidates, and even methods of campaigning. A final, critical ingredient is the candidates who seek the presidential nomination. The parties select a presidential nominee, but party insiders, activists, groups, and identifiers choose among the candidates who enter the race.

Notes

1 D. Jason Berggren, 2007, "Two Parties, Two Types of Nominees, Two Paths to Winning a Presidential Nomination, 1972–2004," *Presidential Studies Quarterly*, 37(2): 203–227.
2 Key, 1959, "Secular Realignment and the Party System."
3 Burnham, 1970, *Critical Elections and the Mainsprings of American Politics*; Sundquist, 1983, *Dynamics of the Party System*; Edward G. Carmines and James A. Stimson, 1989, *Issue Evolution: Race and the Transformation of American Politics*, Princeton, NJ: Princeton University Press.
4 Skowronek, 1993, *The Politics Presidents Make*; Karol, 2009, *Party Position Change in American Politics*.
5 Noel, 2013, *Political Ideologies and Political Parties*.
6 Mayer, 1996, *The Divided Democrats*.
7 Keith Poole, Nolan M. McCarty, and Howard Rosenthal, 1997, *Income Redistribution and the Realignment of American Politics*, Washington DC: AEI Press; Jeffrey M. Stonecash, Mark D. Brewer, and Mack D. Mariani, 2003, *Diverging Parties: Social Change, Realignment, and Party Polarization*, Boulder, CO: Westview Press.
8 The following discussion builds on Paulson, 2007, *Electoral Realignment*; Jeffrey M. Stonecash, 2006, *Political Parties Matter: Realignment and the Return of Partisan Voting*, Boulder, CO: Lynne Rienner; Jeffrey M. Stonecash, 2013, *Understanding American Political Parties, Democratic Ideals, Political Uncertainty, and Strategic Positioning*, New York: Routledge; and Hans Noel, 2013, "Coalition-Building and Ideology in 2012 and Beyond," in *Winning the Presidency*, William J. Crotty (ed.), Boulder, CO: Paradigm, 103–115.
9 Stonecash, Brewer, and Mariani, 2003, *Diverging Parties*.
10 Paulson, 2007, *Electoral Realignment and the Outlook for American Democracy*.
11 Mayer, 1996, *The Divided Democrats*.
12 Greg Adams, 1997, "Abortion: Evidence of Issue Evolution," *American Journal of Political Science*, 41(3): 718–737.
13 Keeter, Funk, and Kennedy, 2005, "Deaniacs and Democrats."
14 "Beyond Red Vs. Blue: the Political Typology," Pew Center Research, Washington DC (June 26, 2014), www.people-press.org/files/2014/06/6-26-14-Political-Typology-release1.pdf.
15 Christine B. Williams and Girish J. Gulati, 2008, "What Is a Social Network Worth? Facebook and Vote Share in the 2008 Presidential Primaries," Paper presented at the annual meeting of the American Political Science Association, Boston, MA.
16 Cohen et al., 2008, *The Party Decides*.
17 Mayer, 1996, *The Divided Democrats*.

9

CANDIDATES

Opportunism, Competition, and Change

Whether party stakeholders can unify before the caucuses and primaries depends in no small part on the candidates who enter the race. Candidates are active agents in the coalition-building process. Ambitious politicians who want to be president look for circumstances in which their chances for success are good, but they also try to improve their chances through innovative strategies, persuasion, and even changing their party coalitions. Nominations tend to be more *candidate-centric* during time periods when party coalitions are fragmenting and unstable and there is less polarization between the parties, as was the case in the late 1960s and 1970s when the major parties were in the midst of a realignment. Nominations in these circumstances also involve more uncertainty about which candidate will be nominated, which leads more strong candidates to enter the race since more of them are inclined to think that they can win. As a result, nominations during periods of party coalition instability will be more competitive with less coalescence among party constituencies during the invisible primary. Presidential candidates also have greater potential to influence the formation of a new party coalition.

"Who runs" matters in a presidential nomination campaign—even when the party coalitions are more stable as they have become over the past twenty or so years. Every candidate brings a unique package of personal characteristics, policy positions, and ideological image to a campaign. Some candidates will have broader appeal among the different constituencies and groups that make up the political parties than do other candidates. While all candidates think that they have the right stuff to be president, their self-image and policy vision may not be shared by the constituencies of their party. Variation in candidate characteristics and appeal affect how readily party stakeholders and identifiers will coalesce behind a candidate, which in turn affects how competitive the campaign will be during the invisible primary through the nominating elections.

Who Runs?

The candidates who have the best chances of becoming the nominee are "traditional" candidates.[1] Traditional candidates hold or previously have held a high-profile elective office. While many high-profile politicians could run, most do not even if they might want to be president. A critical part of understanding "who runs" is that the politicians who make the strongest candidates are strategic about their willingness to run. Strategic politicians are ambitious and opportunistic.[2] Political ambition usually is demonstrated through a career path of seeking a more prominent elected office and then using that office as a platform for rising to an even higher office.[3] Most vice presidents, governors, and senators have advanced to their position through lower level offices. Strategic politicians calculate their chances of winning and they run when they believe those odds are high. They gather information about their chances when deciding whether to run for the presidency. They pay attention to polls. They consult their friends and associates. They may even begin fundraising through a leadership PAC to gauge their ability to raise money for a run for the presidency. They often travel to Iowa and New Hampshire to test the waters.

Strategic politicians may forego a run for the presidency if they don't think the window of opportunity is big enough. Strategic politicians generally avoid harming their current careers so they tend to stay out of a race when they think they are probably going to lose. The most important factor in this calculus is who else is running and how much appeal these rivals have with various party constituencies. Potential candidates tend to be dissuaded from running for president by the presence of a popular candidate in the race who has appeal across the various factions of a party.[4] The effect is self-reinforcing. If a popular candidate enters the race, other strong prospects tend not to run—leaving the strong candidate running against a weaker field of candidates who they can easily beat. If a popular candidate does not run, then the other strong prospects are more likely to run and the race is more competitive with several candidates who will be viable.

By far the strongest candidates are incumbent presidents. Every incumbent president who has sought renomination has won since 1884. There have been a couple of presidents who decided not to run such as Lyndon Johnson in 1968, but even Johnson probably would have prevailed. In that year, Democrats nominated Johnson's vice president who had pledged to continue Johnson's policies. Incumbent presidents are extremely likely to win even if they face a serious challenge from within their party.

An incumbent president or a vice president may be challenged if they are perceived to be unpopular with substantial numbers of party activists and there are major divisions in the party.[5] Ronald Reagan challenged President Gerald Ford for the 1976 Republican presidential nomination. The circumstances, however, were truly exceptional in the history of the United States. Ford had been appointed to the vice presidency and then assumed the Office of the President when President Richard Nixon resigned. Ford also was unpopular after he pardoned Richard Nixon. That combination of atypical events gave Reagan reason to believe that he could win. Presidents Jimmy Carter and George H. W. Bush also faced challenges because they were perceived as unpopular with the substantial segments of the liberal and conservative wings of their respective political parties. Carter was a moderate Democrat who was not popular with the dominant liberal wing of the Democratic Party. As president, Carter had numerous fights with members of his own party in Congress. He was not popular generally, largely because of a weak economy and perceived weakness in foreign policy. Carter also governed at a time when the Democratic Party was highly divided internally. Under these circumstances, it becomes conceivable that a popular politician like Senator Ted Kennedy could have beaten Carter in 1980. Being unpopular, by itself, is not sufficient to deny an incumbent the renomination by his party. The incumbent's party coalition also must have internal divisions. That Ford and Carter prevailed is a testament to the power of incumbency in seeking renomination.

Other presidents and vice presidents have faced less, if any, opposition in their nomination campaigns. A big reason is that the strongest of their potential rivals do not run. Current and former vice presidents have had the best odds of winning an open nomination of their political party in the last century. In that time, every incumbent vice president has won the nomination of his party when he has sought it. Even former vice presidents—those who ran after the opposition party won the presidency—have been more successful than candidates from other offices (see Table 9.1).

Current and former governors and senators also may make the top tier of presidential aspirants. Governors have had executive experience and may have developed networks of potential supporters on a national scale, e.g., Bill Clinton in 1992 or George W. Bush in 2000. The U.S. Senate is often thought of as a presidential incubator. The Senate produces a lot of candidates but only a few of them have been nominated and even fewer have been elected president in the last century.[6] Candidates from the House of Representatives generally lack the name

Table 9.1 Backgrounds of Major Candidates Who Ran in an Open Presidential Nomination, 1972 to 2012

Position	Ran	Won	Success Rate
Vice President	4	3	75.0%
Governor	31	6	19.4%
Senator	56	5	8.9%
U.S. Representative	20	0	0
Major City Mayor	3	0	0
Administration Position	8	0	0
General	1	0	0
Activist	8	0	0
Business	4	0	0

Note: Major candidates are defined as candidates who declared for the office and who were included in a national Gallup poll or who received at least 0.1% of the vote across all of the primaries in a given year. Position is the most prestigious held and most recent position prior to seeking the nomination. Alan Keyes and Pat Buchanan are classified as activists, though both held positions in presidential administrations in earlier decades.

recognition and record to be taken seriously, though party leaders in that chamber may be taken seriously as a presidential candidate such as former Speaker of the House Newt Gingrich in 2012.

Table 9.1 presents the most recent leadership position of all of the major candidates in the modern era (1972 to 2012). Both vice presidents who sought the nomination—George H. W. Bush and Al Gore—have won. Two former vice presidents have sought the nomination, with Walter Mondale winning the Democratic nomination in 1984 and Dan Quayle dropping out of the Republican race in 1996. Six of the 31 governors and former governors who ran have won since 1972. Senators and former senators are by far the largest category of candidates. As one journalist put it, "Sooner or later, senators look in a mirror and see a President."[7] Fifty-six of the 133 major candidates to seek the nomination since 1972 have been a U.S. senator or a former senator. Senators, however, generally do not win the nomination and only three have been nominated since 1972. Current and former members of the U.S. House of Representatives also run frequently, but do not win. Officials in previous presidential administrations also sometimes seek the nomination, though these candidates also have failed in every attempt. Political activists and businessmen also sometimes have sought the nomination. These candidates typically run to raise the visibility of an issue or cause.

These kinds of advocacy candidates have become much more numerous in the post-reform era. While they have appeal with some subset of their political party, these kinds of candidates usually do not make the top tier of candidates considered to have a serious chance of gaining the nomination. They do, however, have the potential to draw enough support to keep the nomination in doubt later into the caucus and primary season. They also may force the major candidates to pay attention to certain issues. These non-traditional candidates generally run to draw attention to a problem such as racial inequities, as Jesse Jackson and Al Sharpton did with their campaigns. On the Republican side, advocacy candidates have run to raise the visibility of a set of policies related to the theme of "family values" as Pat Robertson, Pat Buchanan, and Gary Bauer did in the 1988 to 2000 campaigns. Non-traditional candidates typically run low budget campaigns, structure their campaign to attract media attention, and they usually continue in the race throughout the primaries even when they have no realistic chance of winning. Traditional candidates differ in this respect because they usually withdraw from the race once it becomes apparent that they are unlikely to win the nomination.[8]

A major difference that has occurred since the beginning of the modern era of nominations is that traditional candidates increasingly withdraw from the race if they do poorly in the Iowa caucus or New Hampshire primary. Democratic primaries, in particular, have become less competitive because most U.S. senators, governors, and U.S. representatives have dropped out of the race within two weeks of the Iowa caucus and New Hampshire primary if they don't finish in the top three of these races. Voters in Iowa and New Hampshire are the only voters to face a full slate of active candidates. Voters in states holding primaries later select from a greatly reduced field of candidates. It should be noted that many of these candidates remain on the ballots in these later primaries, essentially providing voters with a symbolic option of someone other than the inevitable nominee. About a third of Republican voters in the later 2012 primaries voted for a candidate other than Mitt Romney even though their choices had dropped out of the race.

Who seeks (and does not seek) the nomination has major consequences for the competitiveness of the race, the outcome of the campaign, and the ideological direction of a political party. Who decides to run (or not run) affects the decisions of other potential candidates and how the field of candidates shapes up. Candidates vary in their appeal to party constituencies on the basis of their personal character (e.g., charisma, competency, integrity, and leadership) as well as issue or policy positions. Some candidates more readily appeal to the various segments of a political party's membership compared to others. The absence of a candidate who has known appeal with the various constituencies of a political party makes it more difficult for party insiders and activists to come together in support of a given candidate.

There are differences in campaign dynamics and outcomes for different kinds of candidate fields.[9] In particular, the race-entry decision of the early favorite—the candidate leading in national preference polls two years prior to the primaries—affects nominations in several ways. First, a decision to run by the early favorite may affect the decisions of others considering a run. If, by running, the early favorite deters just one or two potentially strong candidates, then the early favorite is left competing against weaker candidates who were willing to take the chance that the early favorite stumbles. A decision not to run by the early favorite opens up the race for other candidates, making the race more competitive with more evenly matched candidates going into the primaries.

Second, early favorites tend to gain the resources, exposure, and support needed to run strong campaigns. As Keech and Matthews put it, "anticipations of victory stimulate a flow of publicity, money, experienced staff and other resources toward the probable winner."[10] For example, going into the 2000 Republican nomination campaign, candidates like Senator Lamar Alexander were unable to raise much money. Most of the big fundraisers and donors in the Republican Party were holding their contributions until George W. Bush entered the race. Without an early favorite, resources tend to be distributed more evenly among the candidates. As we saw earlier, the competition for money is especially balanced in Democratic nomination races, which occurs mainly because the early favorite usually does not enter the race.

Third, early favorites tend to do well because they are relatively well-known and have established images, which give them two advantages. Well-known candidates need fewer resources to establish themselves among voters, which enables them to conserve resources for the primary season. Their support also tends to be stable compared to lesser-known candidates. There is remarkable stability in polls from the first quarter of the year prior to the primaries through January of the election year for presidential nomination campaigns in which the early favorite is a candidate.[11] With lesser-known candidates who rise in the polls, the sudden scrutiny of intense media coverage can reveal information about a candidate that voters may not like. For example, in the 2012 Republican nomination, Governor Rick Perry rose and fell in national opinion polls quickly. As he rose to the top of the polls, the news media's greater attention to Perry increased public awareness of his tendency to reveal his lack of depth on issues. His support slipped dramatically after a debate in which he could not recall one of the three federal government agencies he would eliminate if he would be elected president. While single events rarely have this much of an impact on a candidate's support, support for candidates can shift as partisan activists and identifiers learn about them. Support for well-known candidates tends not to change much because party activists and identifiers know who they are and whether they like them or not. Opinions about lesser-known candidates can change dramatically as voters learn new things about them.

From the perspective of party insiders and activists, the presence in the race of the early favorite reduces uncertainty. The early favorite in polls will be a viable candidate. An early favorite is generally known to the party membership and has an established image, which enables stakeholders to estimate whether that candidate will play well with their constituents. Stakeholders have a pretty good idea about what they are getting with a candidate sufficiently well-known to lead in national polls two or three years before a presidential election. Without an early favorite in the race, there is greater uncertainty about which candidates will be viable contenders for the nomination and how well each candidate will play with constituents in the general election. In this more uncertain environment, party stakeholders have less incentive to endorse early and they are more likely to divide their support among several candidates.

In nomination campaigns that have a strong candidate who can attract broad support among party members, the nomination is usually determined before the caucuses and primaries begin and voting in these elections can be viewed as a ratification of decisions made informally during the prior year. The 1980, 1988, 1996, 2000, and 2012 Republican nominations and the 1984 and 2000 Democratic nominations illustrate these kinds of races. In each of these cases the front-runner who attracted widespread support in the year leading up to the primaries had been a well-known politician with a prestigious position and who had polled better than any other potential candidate three years out. When this candidate ran, it was fairly straightforward for party stakeholders and the media to figure out which candidate would be the front-runner going into the caucuses and primaries. When party insiders, activists, and groups rally around one candidate during the invisible primary, that candidate can emerge with an insurmountable lead even before the caucuses and primaries begin.

Party insiders, activists, and groups have a harder time figuring out which candidate will be the most viable, electable, and preferable when either the "apparent" front-runner does not run or when more than one strong candidate enters the race. In nomination campaigns without a strong candidate, or when a political party is highly divided, there is usually no candidate who enters the caucuses and primaries as a clear front-runner. The 1972, 1976, 1988, 1992, and 2004 Democratic nominations and the 2008 and 2012 Republican nomination campaigns illustrate this kind of race.[12]

Indeed, one of the greatest differences in the competitiveness of Democratic and Republican nomination races since 1972 has to do with the decision of the early favorite—the candidate leading in national public opinion polls one or two years before the election year. The prospective candidate with the greatest public opinion support two years before the election year did not run in five of seven open Democratic nomination races. Republicans have not had a race like this—the Republican candidate leading in these early polls has run in every nomination campaign in the modern era. This makes it much more likely that Republicans unify earlier during the invisible primary compared to Democrats, who tend to be more undecided and divided about which candidate they should back for the nomination. Democratic presidential nominations have been more competitive and the outcome in doubt.

For example, in 1969 and 1970, the Democratic politician drawing the most support in national Gallup polls was Ted Kennedy. Kennedy was the leading candidate for the Democratic presidential nomination in advance of both the 1972 and 1976 election years. His decision not to run in these nomination campaigns left a much greater degree of uncertainty about which of the remaining candidates would be viable, electable, and preferable. The resulting races were the most competitive of any on the Democratic side in the post-reform era. The early Democratic front-runner—former Senator Gary Hart—ran in 1988 but he withdrew from the race after getting caught with a model on a yacht named *Monkey Business*. While Hart reentered the race just before the Iowa caucus, his scandal and lack of campaign resources could not be overcome.

Democrats faced an even more uncertain environment in 1991 when none of the party's heavy hitters sought the nomination. Senator Al Gore and House Majority Leader Dick Gephardt both decided early on that they would not seek the nomination. New York Governor Mario Cuomo, who was heavily favored in polls of Democratic Party identifiers, kept equivocating about whether he would run. He did not formally announce that he was not a candidate until November 1991—just a couple months before the Iowa caucus. Cuomo's equivocation froze the race. Many party insiders and activists refused to commit to a candidate while waiting for Cuomo to announce. By the time Cuomo decided not to run, it was too late for other heavy hitters to enter the race. It was also too late for Democratic stakeholders to coordinate their efforts in support of one of the candidates who did run. The vast majority of Democratic Party insiders failed to make an endorsement of a candidate prior to the Iowa caucus in that year. The resulting uncertainty about the race ensured that campaign momentum during the caucuses and primaries would be decisive in determining the nominee.

Hillary Clinton was the early front-runner in national Gallup polls in 2001 and 2002. Her decision not to run for the presidency in 2004 similarly contributed to a more open candidate field. A relatively unknown governor from Vermont, Howard Dean, emerged as an early front-runner in 2003, but he never had a strong lead in polls and he lacked the support of most elite Democratic Party insiders. Dean's campaign began to slide about 10 days before the Iowa caucus when he was upset by Senator John Kerry, who was not an accomplished senator. Kerry, however, had the benefit of having served during Vietnam and was widely viewed as being a more credible candidate to campaign against George W. Bush on the Iraq War.

In general, campaigns are less competitive and have less uncertainty when a popular politician with national name recognition decides to enter the race. That candidate tends to attract the most support among party insiders and activists who rally to that candidate during the invisible primary. That candidate usually raises the most money and gains the most favorable news coverage. And that candidate usually wins the nomination during the caucuses and primaries. Campaigns become highly competitive and there is a great deal of doubt when either the early front-runner does not enter (or drops out of) the race, as happened for Democrats in 1972, 1976, 1988, 1992, and 2004. Races are more competitive when several strong candidates enter the race, as happened for Republicans in 1988 when Vice President Bush and Senate Minority Leader (and former vice presidential candidate) Bob Dole sought the nomination. In these races, party insiders tend to refrain from declaring their support or they divide their support among the candidates. Party identifiers also tend to divide as evidenced by polls and news coverage of these campaigns. These nominations tend to be influenced by campaign momentum during the primaries, as we will see in the next chapter.

Candidate Opportunism as a Driver of Change

Political opportunism has other implications for presidential nomination campaigns because it affects the strategy and behavior of candidates. Opportunistic candidates seek to take advantage of circumstances and they try to improve their chances of success. Politicians who are ambitious and opportunistic try to gain the support of individuals and groups who can help advance their cause. In particular, some candidates who lacked support from the more mainstream or "establishment" faction of the Democratic and Republican Parties have sought to improve their chances of success by bringing into the party new voters in the primaries or by activating those individuals and groups that might otherwise sit out a nomination. This matters because who participates affects the nomination and ultimately the ideological direction of the party. Policy activists are drawn to candidates who promise to champion their causes. Issue activists have used candidates' campaign organizations to become part of the party coalitions.[13] Whether party activists are veterans or newcomers to the process, support for particular candidates plays an important role in shaping the parties' ideological orientations.[14]

Candidates appeal to policy seeking activists to expand their own base of support within the party in order to improve their chances of getting nominated. Candidates also may seek to alter their party's commitment to certain policies by motivating greater participation by certain groups within the party's constellation of constituencies. Both goals may lead candidates to mobilize specific constituency groups in their party.

Candidates themselves may seek to bring in new constituencies to the party by promoting an ideological vision of public policy. Ronald Reagan used issues like abortion, gun control, and crime to expand the appeal of the Republican ticket among white evangelical Christians. Reagan was ahead of the Republican Party in this regard. His campaign worked with representatives of the Moral Majority, a group founded by Reverend Jerry Falwell in 1976, to bring white evangelical Christian activists into the Republican Party through the 1980 nomination campaign. Abortion was an issue that cut across political party cleavages in the 1970s; both Republicans and Democrats in the U.S. Senate were about as likely to take pro-choice as pro-life positions.[15] Reagan's appeal to social conservatives helped make this group a central part of the Republican Party's electoral base. In doing so, Reagan played a critical role in redefining the ideological orientation of the Republican Party.

Candidate efforts to redefine their political party and draw in new voters are not limited to Republicans. Liberal Democrats like George McGovern and Ted Kennedy explicitly attached themselves to the liberal civil rights movements and expanded Democratic Party appeal to African Americans, single women, Latinos, and LGBT populations. These candidates drew civil rights activists and social liberals into the nomination process of the Democratic Party. Presidential candidates thus became one of the vehicles by which the Democratic Party became distinctly more liberal on social issues. While the liberals of the civil rights movement became integrated into the Democratic Party and "movement conservatives" became integrated into the Republican Party, it is candidates who help galvanize these movements and give them direction within the framework of a political party. Thus candidates and party constituencies interactively influence the policy direction of their party's nominations for elective office.

Short of having a realistic chance of getting nominated, candidates might appeal to activists as a means of self-promotion or to enhance the party's commitment to certain policies. For example, the Reverend Al Sharpton's campaign for the 2004 Democratic nomination arguably was a bid by Sharpton to supplant the Reverend Jesse Jackson as the most recognized national spokesperson of African Americans. Jackson had emerged as the most visible African American voice on politics and society after his 1984 and 1988 campaigns for the Democratic presidential nominations. Jackson's Rainbow Push Coalition retains a substantial organizational presence and continues to influence education policy in a number of states. Both campaigns sought to raise the visibility and salience of issues concerning African Americans, even if the candidates were not likely to win the nomination.

The Reverend Pat Robertson's 1988 Republican nomination campaign could be viewed similarly. Although he did not have a realistic chance of winning the Republican presidential nomination in 1988, Robertson's presidential nomination campaign expanded the voice of evangelical Christians in the Republican Party. Robertson created the "Christian Coalition" as a political organization in that same year. The Christian Coalition replaced the Moral Majority, which had been established in 1976 by Jerry Falwell. Once integrated into political party networks, these party activists demand that candidates adhere to positions. Thus by 1996, prominent Republican presidential aspirants who are pro-choice, like California Governor Pete Wilson, no longer had a realistic chance of winning the party's nomination.

The nomination of a president is not just about selecting a candidate. It is also a competition among candidates and party activists for the definition of the policies advanced by a political party. As governor of Arkansas, Bill Clinton had been a founding leader of the Democratic Leadership Council (DLC), a group of moderate Democrats who formed a group within the Democratic Party seeking to advance a more fiscally conservative version of liberalism, summarized by Clinton's prescription that "government should do more with less." The DLC was formed to advance a moderate position in order to increase the appeal of the Democratic Party among Southern voters and especially suburban voters. About half of all Americans live in suburbs of metropolitan areas. Expanding the Democratic Party's appeal among suburbanites was considered critical to revitalizing the Democratic Party as inner cities continued to lose population to the suburbs. Bill Clinton and his Vice President Al Gore were both DLC members, who sometimes disagreed with liberals in their party. The DLC–Liberal factionalism of the Democratic Party played into the 2008 nomination campaign. Although both Barack Obama and Hillary Clinton were DLC members, most elite elected officials who identified with the DLC backed Clinton while more liberal officials supported Obama.

There are limits on what candidates can do to change their party and still have a chance to win a nomination. The coalition-building process is interactive and party activists demand that candidates adhere to preferred policy positions. For example, Democrats Ted Kennedy and Dennis Kucinich had taken the pro-life position on roll call votes in Congress prior to their running for the Democratic presidential nomination. Both adjusted their positions to fit with the dominant position of Democratic Party activists in order to increase their appeal among these critical constituencies. Similarly Republicans Bob Dole and John McCain had long records of fiscal conservatism in Congress. Fiscal conservatism is an economic philosophy that prescribes balanced budgets. While all Republican candidates profess to want balanced budgets and smaller government, fiscal conservatives are willing to raise taxes to balance budgets. Supply-side Republicans have sought tax cuts even if it means bigger government deficits. Former Vice President Dick Cheney chided skeptics, saying that "Reagan proved deficits don't matter."[16] What mattered more was getting tax cuts for a major party constituency. Dole in 1996 and McCain in 2008 promised to cut taxes, consistent with the supply-side views that had become prevalent among Republican Party activists by that time.

The interaction between opportunistic politicians and policy-demanding activists has raised the importance of authenticity and integrity as personal characteristics of successful presidential aspirants. Political activists look to find a credible champion of their causes, and they are skeptical of aspirants who change positions or appear to be pandering. Nearly all presidential candidates promise to support policies preferred by key party constituencies, but not all are viewed as credible champions of those policies. Some presidential candidates may have difficulty attracting the support of some political party constituencies that have doubts about the candidate's commitment to certain policy positions. For example, some Republican activists who lauded Ronald Reagan's tax policies in the 1980s were dubious of George H. W. Bush's commitment to that position. As a presidential candidate in 1980, Bush had mocked Reagan's policies as "voodoo economics." Similarly, supply-side conservatives in the Republican Party were dubious of John McCain's embrace of tax cuts in 2008 after he had opposed George W. Bush's tax cuts in 2001 and 2003. Mitt Romney was viewed warily by many movement conservatives during the nomination campaign of 2012. Romney was a Mormon who had been governor of a liberal state that had enacted a healthcare program similar to the Affordable Care Act (aka Obamacare). Romney also had expressed support for gay rights when he ran for the U.S. Senate in 1996. All of these candidates had to pledge their commitment to the dominant party positions on these policies in order to win enough support to be nominated. While conservative Republicans may have had doubts about George H. W. Bush, John McCain, or Mitt Romney's commitment to particular policies, conservative Republicans supported these nominees in the general election when the alternative was a Democrat.

Conclusions

The strongest presidential candidates are politicians in high prestige public offices, who have national name recognition, and who attract support in public opinion polls taken years before the caucuses and primaries. These politicians tend to be strategic and opportunistic. They try to figure out their chances of winning and they are more willing to enter the campaign if they estimate that they have a good chance of winning the nomination and general election. The decision of these candidates to enter the race or to remain on the sidelines has a considerable effect on the competitiveness of a presidential nomination campaign. Other strategic politicians may be deterred from running in a presidential nomination campaign that has as a candidate a particularly popular politician. That makes it relatively easy for party insiders and activists to figure out who they should support for the nomination. This scenario tends to have party stakeholders endorsing one candidate disproportionately more than other candidates and the "anointed" front-runner receives a disproportionate share of media coverage, fundraising, and support in public opinion polls of party identifiers. That is, the party insiders and activists decide who should be the nominee during the invisible primary, but they choose among the candidates in the race. When the early front-runner is one of the candidates, then the various constituencies of the political party unify or coalesce around the front-runner.

When the early front-runner decides to stay on the sidelines, as Senator Ted Kennedy did in the 1970s or Mario Cuomo did in 1988 and 1992, then the race is relatively more competitive. There is a similarly competitive race when two or more strong candidates enter the nomination campaign, as Vice President George H. W. Bush and Senate Minority Leader Bob Dole did in the 1988 Republican campaign. In both of these situations, party stakeholders tend to take a wait-and-see attitude. They refrain from publically committing to a candidate, waiting to see which candidate will emerge as relatively more popular with party activists and identifiers. The lack of insider and group signals tends to add to a campaign that is more uncertain and competitive. Without a strong or clear signal about who they should support, party activists and identifiers tend to divide among the candidates and no candidate emerges as a strong front-runner before the caucuses and primaries. Public opinion polls tend to show considerable instability in these nomination campaigns as different candidates may rise to the top of the polls and absorb considerable media scrutiny. A lot of candidates cannot withstand that scrutiny and their public support wanes. In these nomination campaigns there is a tendency for the winner to be the candidate who emerges during the caucuses and primaries.

Political party insiders, activists, and groups select their nominee but they choose among the candidates in the race. For their part, the strongest candidates have developed their reputations, honed their skills, cultivated relationships, and built a national network of supporters for years before they decide to run. These candidates are opportunistic in that they will take advantage of favorable circumstances to run but also in that they will try to build a winning coalition of people and groups who may or may not be active participants in presidential nominations before the candidate's campaign.

A candidate's ability to appeal to and draw the support of party insiders, activists, and groups depends on their ability to fit the expectations and demands of these partisans. They have to have a reputation for policy that fits with these expectations if they are to be credible champions of policy. Candidates begin positioning themselves for a nomination campaign years before they run. John McCain, for example, began to deviate in his voting record more during the four years leading up to his first presidential campaign in 2000 than he had in prior years in the Senate.[17] McCain continued his "deviant" voting record through Bush's first term but returned to a more orthodox voting pattern in the four years leading up to the 2008 election when he did become the Republican nominee.

Candidates' personal characteristics—integrity, competence, charisma, empathy, and authenticity—affect their ability to resonate with party constituencies. Party activists want a champion but that champion cannot be a dud on the campaign trail. Which characteristics are important in a given year seems to vary across elections. Which characteristics are emphasized depends on which candidates are in the race and what kinds of questions have emerged with the incumbent president's party. Integrity, for example, was a major topic in the 1976 Democratic nomination, when Jimmy Carter ran on his integrity as much as anything in the aftermath of the Watergate scandal that forced Richard M. Nixon to resign the presidency. Carter had a contentious relationship with his own political party as president, and leadership became a major topic in the 1980 Democratic and Republican nomination campaigns. Ultimately what matters is how persuasive a candidate is when he or she appeals to party insiders, activists, and groups aligned with the political parties. Successful candidates demonstrate an ability to appeal to and persuade party insiders, activists, and aligned groups to support them.

Candidates thus play a critical role in nomination campaigns as active participants in the coalition-building process. Political parties are broad coalitions, and who actively participates in the presidential nomination campaign varies somewhat in each election year. Candidates have the potential to influence the constituencies and policies of their political party by affecting who participates in the selection of the presidential campaign.

Notes

1 Norrander, 2006, "The Attrition Game."
2 Adkins, Dowdle, and Steger, 2006, "Progressive Ambition, Opportunism, and the Presidency"; Brown, 2011, *Jockeying for the American Presidency*.

3 Joseph Schlesinger, 1966, *Ambition and Politics: Political Careers in the United States*, Chicago: Rand McNally; Peabody, Ornstein, and Rohde, 1976, "The United States Senate as a Presidential Incubator."

4 Randall E. Adkins, Andrew J. Dowdle, Greg Petrow, and Wayne Steger, 2015, "Ambition, Opportunism, and the Presidency, 1972–2012," Paper presented at the annual meeting of the Midwest Political Science Association, Chicago.

5 Steger, 2003, "Presidential Renomination Challenges in the 20th Century."

6 Barry C. Burden, 2002, "United States Senators as Presidential Candidates," *Political Science Quarterly*, 117(1): 81–102; Wayne P. Steger, 2006, "Stepping Stone to the White House or Tombstone on Presidential Ambition? Why Senators Usually Fail When They Run for the White House," *American Review of Politics*, 27(1): 45–70.

7 Robin Toner, 1991, "The Field Is Still Open, Time Wanes, and, as in '88, the Name Gore Arises," *New York Times*, March 25.

8 Norrander, 2006, "The Attrition Game."

9 Butler, 2004, *Claiming the Mantle;* Keech and Matthews, 1976, *The Party's Choice*; Berggren, 2007, "Two Parties, Two Types of Nominees, Two Paths to Winning a Presidential Nomination."

10 Keech and Matthews, 1976, *The Party's Choice*, 53.

11 Andrew J. Dowdle, Randall E. Adkins, and Wayne P. Steger, 2009, "The Viability Primary: Modeling Candidate Support Before the Primaries?" *Political Research Quarterly*, 62(1): 77–91.

12 Adkins et al., 2015, " Ambition, Opportunism, and the Presidency, 1972–2012."

13 Herrera, 1995, "The Crosswinds of Change," 293.

14 Herrera, 1995, "The Crosswinds of Change."

15 Adams, 1997, "Abortion: Evidence of Issue Evolution."

16 Ron Suskind, 2004, *The Price of Loyalty: George W. Bush, the White House, and the Education of Paul O'Neill*, New York: Simon & Schuster.

17 McCain's DNominate Scores begin to drift more in the late 1990s than they had earlier when McCain had established a very conservative record. See http://voteview.com/dwnominate.asp for more details on the scores.

10

BEFORE, DURING, AND AFTER
THE PRIMARIES

Political scientists who study presidential nominations essentially offer two frameworks for explaining the presidential nomination process in the modern era. The explanations that emerged in the 1970s and 1980s often focused on whether candidates gained or lost momentum during the caucuses and primaries.[1] These explanations can be considered to be *candidate-centric* in that they focus on various aspects of candidates' campaigns and on the candidates' appeal with party identifiers. Candidates' policy positions and personal characteristics like leadership, charisma, and integrity are central factors in candidate appeal. Candidates' abilities to communicate to voters—through organizational outreach, advertising, and the mass media—also affect their chances of winning a nomination. Campaign finance matters because money is needed to pay for professional staff and advertising. A candidate's chances of winning ultimately depend on their appeal to voters in primaries and caucuses.

The earliest caucuses and primaries have a substantial impact because these contests winnow the field of candidates. Voters in Iowa and New Hampshire do not necessarily pick the nominee, but they certainly contribute to candidates dropping out of the race. Candidates who fail to do well in the earliest caucuses and primaries receive less media coverage, which tends to be even more critical; their fundraising dries up; and they are unable to compete in enough caucuses and primaries to get the convention delegates needed to win the nomination. Candidates who finish in the top two in the earliest contests gain momentum, and one of these two candidates has been able to continue to grow their support in subsequent caucuses and primaries, eventually gaining enough delegates to become the nominee.

An alternate, party-centric perspective holds that presidential nominations are essentially determined before the caucuses and primaries— during the invisible primary—through coordination and collusion among party insiders, activists, and the leaders of groups affiliated with the parties.[2] In this narrative, party stakeholders evaluate candidates and signal each other about which candidate they prefer and will support. Stakeholders have an incentive to coordinate their efforts in order to ensure that the nominee is committed to defending key policy positions that are near and dear to their hearts. Party stakeholders facilitate a preferred candidate's efforts to become the nominee by talking up the candidate, critiquing other candidates, and helping the candidate's fundraising, organizing, and media efforts to persuade rank-and-file partisans. Those rank-and-file partisans learn which candidate they should support from the cues given by party stakeholders. Campaign momentum during the primaries matters less in this narrative because the nominee is largely determined during the invisible primary.

While these explanations tend to focus on different sets of factors, both explanations contribute to our understanding of presidential nominations. Both sets of factors operate in every nomination, though some presidential nominations appear to be more party-centric while other nominations operate more along the lines of the candidate-centric explanation. The difference depends on whether and to what extent party stakeholders are able to unify behind a candidate during the invisible primary. Stakeholders are able to coalesce behind a candidate to a greater degree in some nomination campaigns while in other campaigns many stakeholders remain undecided about which candidate is the best or they divide their support.

Democrats were divided from the 1960s through the 1990s, though they appear to be gaining unity in the last decade.[3] Republicans were relatively divided in the 1950s to the 1970s with a moderate faction represented by Dwight D. Eisenhower and Nelson Rockefeller and a conservative faction most clearly represented by Barry Goldwater. The nomination of Ronald Reagan in 1980 seems to have produced a fairly unified coalition that remained so through George W. Bush's presidency. Republican nominations in 2008 and 2012 suggest that the Republican Party's unity is starting to fray.

Both the availability of candidates and the unity of the party coalitions affect whether and to what extent party stakeholders will coalesce behind a candidate during the invisible primary. Political party coalitions generally are stable, holding together for decades. The groups that form the core coalition of a political party resist change.[4] Party coalitions, however, do shift as a result of long-term changes in the membership composition of the parties. Nominations are also affected by short-term fluctuations in participation by party members. Participation varies mainly with changes in the set of issues that are salient in a given election year and with the candidates.[5] Variations in participation matter because it is the mix of citizens who vote that select the nominee.

Whether and the extent to which stakeholders coalesce behind a candidate is affected by which candidates seek the nomination. Party stakeholders and primary voters have to choose among the candidates who will be on the ballot—not some mythical ideal candidate. Who decides to run (or not run) affects the decisions of other potential candidates and how the field of candidates shapes up. Presidents, vice presidents, and nationally known politicians tend to make the strongest candidates. The presence of one of these candidates usually deters other strong candidates from entering the race—making it easier for party stakeholders and identifiers to unify behind a frontrunner. Candidates vary in their appeal to party constituencies on the basis of their personal character (e.g., charisma, competency, integrity, and leadership) as well as issue or policy positions. Some candidates more readily appeal to the various segments of a political party's membership compared to others. Candidates like Ronald Reagan and George W. Bush had strong appeal to economic conservatives as well as to social conservatives. When politicians like these enter the race, the decision about which candidate to support is much easier for the myriad groups that form the national political parties. The absence of a candidate who can appeal to the various factions of a political party makes it more difficult for party insiders and activists to come together in support of a given candidate.

In nomination campaigns that have a candidate who can attract broad support among party members, the nomination is usually determined before the caucuses and primaries begin and voting in these elections can be viewed as a ratification of decisions made informally during the prior year. There was little doubt in 1979 that Ronald Reagan would be the Republican presidential candidate in 1980. Republicans had a good idea that Vice President George H. W. Bush would be the nominee in 1988, that Senate Majority Leader Bob Dole would be the nominee in 1996, and that Texas Governor George W. Bush would be the nominee in 2000. Democrats have had races in which the presidential nominee was known by the end of the invisible primary, including in 1984 when former Vice President Walter Mondale ran in the race with disproportionate backing of party insiders and groups. There was little question that Vice President Al Gore would be the Democratic nominee in 2000. In every one of these campaigns, the news media continued to report as if the horse race was close and that the outcome was in doubt, but these news stories often took the form of speculative narrative along the lines of "if the front-runner stumbles, then . . ." But not all presidential nominations look like this. Party insiders and activists sometimes fail to coalesce behind a candidate during the invisible primary, a fact that indicates a limitation of the "invisible primary" explanation of presidential nominations. There is more uncertainty about which candidate will become the presidential nominee of a political party as stakeholders divide their support or refrain from committing to a candidate. Uncertainty itself affects the decisions of potential candidates, insiders, activists, and the media, ultimately reinforcing the competitiveness of a race. When stakeholders remain divided or undecided, the nomination race remains competitive through at least the early caucuses and primaries which winnow the field of candidates.[6]

In nominations when party insiders, activists, and groups fail to unify or refrain from supporting candidates, no single candidate emerges as the clear front-runner before the primaries. The 1988, 1992, 2004, and 2008 Democratic nominations and the 2008 and 2012 Republican nomination campaigns illustrate this kind of race. This usually happens in campaigns without a strong candidate, or when a political party is highly divided. Different party constituencies support different candidates or they remain uncommitted until after the caucuses and primaries begin. Failing to unify before the primaries means that party insiders, activists, and aligned groups deprive rank-and-file partisans the signal needed to figure out which candidate they should support.

In these races, candidates tend to be more evenly matched in their fundraising. The news media also give multiple candidates enough coverage to become known to activists and rank-and-file party members. Public opinion polls may show a leading candidate, but that candidate's support is well below a majority. The 1976 and 1988 Democratic campaigns illustrate this point. In 1976, George Wallace was the leading Democrat in the last national Gallup poll before the Iowa caucus. Wallace had the support of just 18% of respondents—hardly a strong front-runner. In 1988, Jesse Jackson was the leading Democratic candidate with the support of just 22% of respondents in the last national Gallup poll before the Iowa caucus. The 2008 Republican nominee, Senator John McCain, ran third in national Gallup Polls throughout most of the invisible primary while the leader in most national polls was Mayor Rudolph Giuliani who ultimately received very little support from caucus and primary voters. The 2012 Republican nomination had different candidates lead in pre-primary polls, including Romney, Governor Rick Perry, businessman Herman Cain, former Governor Mike

Huckabee (who didn't even run), and former Speaker of the House Newt Gingrich. None received as much as 40% in a given national poll.

Nomination campaigns that are relatively open and competitive when the caucuses and primaries begin are usually determined by campaign "momentum." With less of a guide to which candidate is the best nominee, the earliest caucuses and primaries attract the most competition among candidates. Voters in these states have a disproportionate impact on the nominations.[7] Though these states are often criticized for being unrepresentative of voters nationally, the voters in these states tend to be better informed about the candidates as a result of the intensive retail and wholesale campaigning by the candidates. In most other states, there is less local news coverage, advertising, or candidate appearances until just days before the caucus or primary begins. Voters in the early states like Iowa and New Hampshire, in contrast, are bombarded by months of television and radio ads, daily phone calls, and campaign events. While voters in these states have a disproportionate influence on the outcome, they are in a better position to make informed decisions about the candidates. That the voters who make the most consequential decisions are better informed is probably a good thing for the country from the standpoint of democracy. Governance works best if voters are able to make informed choices. That happens in Iowa and New Hampshire more than in other states.

The ability of voters in Iowa and New Hampshire to make informed decisions is reflected in the candidate preferences of voters in these states—which often tend to be better predictors of the top candidates than are national polls.[8] Voters in New Hampshire had written off candidates like Mayor Rudolph Giuliani and Mike Huckabee well before they faded in national polls in the 2008 Republican nomination. Similarly, New Hampshire voters largely discounted candidates like Rick Perry and Herman Cain even while these candidates were at the top of national polls in 2011.[9] With more information about the candidates, Republicans in New Hampshire were able to figure out that these candidates lacked the characteristics that would enable them to be nominated. Thus having two smaller states have primaries earlier than other states provides an informed filter to winnow candidates. Most of the objections to the privileged status of Iowa and New Hampshire come from the supporters of candidates who lose. The supporters of candidates that win the nomination don't object and these supporters become the convention delegates who have the power to change the nominating process at the conventions. It seems unlikely that there will be a change in the rules of the nominating process in the near future.

Looking Ahead

The arguments of this book should give us a guide to the future if they are correct. As of the time when this book was written, the 2016 presidential election is more than two years away. The arguments of the book apply to the 2016 nominations as follows. First, former Senator and former Secretary of State Hillary Clinton is the front-runner for the Democratic nomination three years out (as of early 2014). If she runs, it should be relatively straightforward for Democratic Party elites, activists, and group leaders to decide whether she would be a good presidential nominee. Given the reasons that she leads polls, a run by Hillary Clinton should enable Democrats to unify well in advance of the caucuses and primaries. If Clinton does not run, however, then it would seem likely that the Democratic nomination would be wide open and a lot more politicians will enter the race in hopes of securing the nomination. In this scenario, it seems unlikely that the 2016 Democratic nomination would be settled before the caucuses and primaries and that momentum during those nominating elections will determine the nominee.

The 2016 Republican nomination is shaping up to be a wide-open race. First, there is no clear front-runner who would unify the Republican Party. While potential candidates like New Jersey Governor Chris Christy, Tennessee Senator Rand Paul, and Florida Senator Marco Rubio are often mentioned as strong candidates, none have the same gravitas and appeal that George W. Bush had among Republicans in the years leading up to the 2000 nomination. In part this reflects the increasingly visible divisions among Republican Party insiders, activists, and group leaders. Even groups affiliated with the Republican Party cannot easily resolve their internal differences to unify on presidential candidates. Given these internal divisions within the Republican Party coalition, it seems reasonable to expect the 2016 Republican nomination to be one in which party insiders and activists fail to unify sufficiently to anoint a candidate as the inevitable nominee. Instead, the nominee will likely be decided during the caucuses and primaries as the party's nominating electorate evaluates the candidates after each successive caucus and primary along the lines of the momentum model described earlier.

Final Reflections

The selection of the nominee matters greatly for the policy or ideological direction and image of a political party. As a result, presidential nominations involve competition and rivalry among party leaders and activists for control of the direction of the parties. Contemporary presidential nominations attract substantial attention and involvement by party elites, activists, and advocacy groups aligned with the political parties. Activists and groups influence nominations through their support for candidates before and during the caucuses and primaries.

The expanded participation in the selection of presidential candidates during the 1970s is consistent with the principle of representative democracy which holds that power should derive from people directly or indirectly. Presidential nominations can be said to be less democratic as party insiders exert more influence over the selection of the nominees as they do when they unify behind a candidate during the invisible primary. This does not mean that party stakeholders have an oligarchical control over the nomination in the way that party bosses may have had a century ago. Rather, party stakeholders' influence is more limited to talking up a candidate, offering assistance in fundraising, and serving as spokespersons for the campaign.[10] Collectively, however, party stakeholders may be able to mediate the nomination by guiding the choices of the larger numbers of party activists and party identifiers who vote in caucuses and primaries. But to do so, they need to be actively involved and unified behind a candidate before the caucuses and primaries begin. Nominations can be said to be more democratic to the extent that the race is competitive and the selection of the nominee depends on the choices of caucus and primary voters. The question is whether these voters have meaningful choices.

In one version of democratic theory, competition among political organizations and leaders is what enables people to make meaningful choices in elections.[11] When a particular candidate has a huge lead because party elites, party activists, and groups have unified behind that candidate, then there is less effective competition among candidates during the caucuses and primaries.[12] During a campaign with an un-level playing field with an imbalance of resources and insider support, voters basically have a vote of confidence (or no confidence) in the candidate preferred by party stakeholders. If party elites, activists, groups, and campaign contributors remain undecided or divide their support among different candidates, however, then caucus and primary voters become the arbiters of the nomination competition. Presidential nomination campaigns can be considered more democratic when caucus and primary voters exert more power over the selection of the eventual nominee. In most respects, it appears that the reforms of the 1970s did make presidential nominations more democratic.

Making participation in presidential nominations more representative of party activists, however, may not have been beneficial for the general population of the United States. Party activists are not representative of the broader population in that they tend to be better educated, a little more affluent, and quite a bit more liberal or conservative than the average citizen. Most citizens are relatively apathetic about politics. Most citizens do not care enough about politics to attend rallies, volunteer, or contribute to candidates. Most people do not vote in the caucuses and primaries that select the convention delegates who nominate presidential candidates. By opening up the presidential nomination process to party activists, the McGovern-Fraser reforms increased the voice and power of policy demanding activists in both political parties. The result may be a nomination process that is more democratic and representative of party constituencies, but that produces outcomes that are less representative of the general public.

Although Americans are becoming more ideologically oriented in their political preferences, many if not most do not have consistent preferences on public policy. Rather, most Americans prefer a mix of liberal and conservative policies. Very often, that mixture of policy preferences ignores the trade-offs that exist in politics. Most Americans, for example, want lower taxes, and they want to maintain or increase spending on government programs like Social Security, Medicare, education, environmental protection, and more. Politicians of both political parties rarely offer stark choices of the trade-offs involved in popular policies. Rather politicians focus on those parts of their policy repertoires that are politically popular and they are silent or ambiguous about the costs of their policies. During the general election, that partial presentation of policy promises focuses on that which is popular to most Americans. During the nomination stage—when the audience is more ideologically motivated with more intense preferences—candidates focus more clearly on what they would hope to do in office. Americans need to pay attention to the politicians at the nominating stage of the election to discover the more stark positions that each political party offers. Americans can make more informed choices on Election Day by paying attention to candidates during the presidential nominating phase of the campaign.

While nominations to federal office have become more democratic, this has not necessarily produced better governance. The growing influence of intensely ideological party activists in the nomination process has affected governance after the elections are over. The growing influence of party activists in the party nominations has contributed to the growing polarization of Congress by promoting increasingly conservative Republican members of Congress and increasingly liberal Democratic members of Congress.[13] The nominees who have been elected president, however, have not become more ideological. Presidents are nominated by national party coalitions. The myriad groups and factions of the national political parties are more diverse than those that exist at the state or even at the regional level. Presidential candidates' campaign speeches and messages are crafted to appeal to party activists, but presidential nominees usually shift toward the ideological center once they are nominated in order to appeal to the even broader set of citizens who will cast ballots in the general election.

Governance has been affected by polarization, but mainly because presidents have to deal with a more ideological Congress.[14] The United States has a constitutional system of separate institutions and checks and balances among the branches of government. In practice, this means that separate institutions share and compete for power.[15] Changing government policy requires agreement among officials in these institutions. When a president's party controls both the House and Senate, we tend to see more partisan legislation being enacted as a result of polarization. The majority takes advantage of political control to implement policies that are desired by party activists. Absent agreement, compromise is generally needed to enact laws. Polarization of the political parties in Congress reduces the room for compromise. When the political parties are highly polarized, as they have become over the past twenty-five years, the result of divided government is a great deal of political posturing, blame games, and little change in public policy. This is a source of frustration to party activists who have very strong preferences for policy. Party activists tie the hands of their elected officials in Congress who are afraid to compromise because they fear a challenge to their own renomination during the next election cycle. Instead of compromise, polarization increases the incentives to posture in office—making for more symbolic politics than substantive policy making. The activists and groups work harder and donate even more money to win elections rather than accept compromises.

In this context, presidential candidates walk a fine line between appealing to ideological party activists during the nomination phase of the election and appealing to a broader set of voters in the general election. As president, their ability to make good on their campaign promises depends mainly on whether their political party controls the majority of seats in the House and Senate. Presidents increasingly represent a divided and angry country in which Democrats deny the legitimacy of

Republican presidents and Republicans deny the legitimacy of Democratic presidents. Many citizens are turned off by the bickering and posturing. The nomination process may be more democratic, but public satisfaction with government has gotten worse, not better.

Notes

1 Aldrich, 1980, *Before the Convention*; Patterson, 1980, *The Mass Media Election*; Bartels, 1988, *Presidential Primaries*; William R. Crotty and John S. Jackson III, 1985, *Presidential Primaries and Nominations*, Washington DC: Congressional Quarterly; Martin P. Wattenberg, 1991, *The Rise of Candidate Centered Politics*, Cambridge, MA: Harvard University Press; William G. Mayer, 1996, "Forecasting Presidential Elections," in *In Pursuit of the White House*, William G. Mayer (ed.), Chatham, NJ: Chatham House, 44–71; Randall E. Adkins and Andrew J. Dowdle, 2000, "Break Out the Mint Julips: Is New Hampshire the Primary Culprit Limiting Presidential Nomination Forecasts?" *American Politics Quarterly*, 28(2): 251–269.
2 Keech and Matthews, 1976, *The Party's Choice*; Cohen et al., 2008, *The Party Decides;* Masket, 2009, *No Middle Ground;* Karol, 2009, *Party Position Change in American Politics;* Noel, 2013, *Political Ideologies and Political Parties*; Bawn et al., 2012, "A Theory of Parties."
3 Mayer, 1996, *Divided Democrats*; James A. Reichly, 2000, *The Life of the Parties: a History of American Political Parties*, Lanham, MD: Rowman & Littlefield.
4 Karol, 2009, *Party Position Change in American Politics*.
5 *Beyond the Red vs. Blue Political Typology*, Pew Research Center.
6 Bartels, 1988, *Presidential Primaries*.
7 Steger, Adkins, and Dowdle, 2004, "The New Hampshire Effect in Presidential Nominations."
8 Dante Scala and Andrew Smith, 2007, "Does the Tail Wag the Dog? Early Presidential Nomination Polling in New Hampshire and the U.S.," *American Review of Politics*, 29(3): 401–424; Dante Scala, 2003, "Re-Reading the Tea Leaves: New Hampshire as a Barometer of Presidential Primary," *PS Political Science & Politics*, 2: 187–192.
9 Smith and Moore, 2015, *Out of the Gate*.
10 Whitby, 2014, *Strategic Decision-Making in Presidential Nominations*.
11 Schattschneider, 1960, *The Semi-Sovereign People;* Schumpeter, 1942, *Capitalism, Socialism, and Democracy*; Held, 1987, *Models of Democracy*.
12 Steger, Hickman, and Yohn, 2002, "Candidate Competition and Attrition in Presidential Primaries."
13 John H. Aldrich and David W. Rhode, 2001, "The Logic of Conditional Party Government: Revisiting the Electoral Connection," in *Congress Reconsidered*, 7th ed., Lawrence C. Dodd and Bruce I. Oppenheimer (eds.), Washington DC: CQ Press, 269–292; Geoffrey C. Layman, Thomas M. Carsey, John C. Green, Richard Herrera, and Rosalyn Cooperman, 2010, "Activists and Conflict Extension in American Party Politics," *American Political Science Review*, 104(2): 324–346.
14 Jeffrey E. Cohen, 2011, "Presidents, Polarization, and Divided Government," *Presidential Studies Quarterly*, 41(3): 504–520.
15 Charles O. Jones, 1994, *The Presidency in a Separated System*, Washington DC: Brookings Institution.

INDEX